SOME RECENT
ADVANCES IN STATISTICS

Based on a Symposium held in 1980 under the sponsorship of the Lisbon Academy of Sciences in their Series: Frontiers of Science.

SOME RECENT ADVANCES IN STATISTICS

edited by

J. Tiago de Oliveira
Academia das Ciências de Lisboa
Rua da Academia das Ciências, Lisboa
Portugal

and

Benjamin Epstein
Faculty of Industrial Engineering and Management
Technion-Israel Institute of Technology
Haifa, Israel

1982

ACADEMIC PRESS

A Subsidiary of Harcourt Brace Jovanovich, Publishers

London   New York
Paris   San Diego   San Francisco   São Paulo
Sydney   Tokyo   Toronto

ACADEMIC PRESS INC. (LONDON) LTD.
24–28 Oval Road
London NW1

United States Edition published by
ACADEMIC PRESS INC.
111 Fifth Avenue
New York, New York 10003

Copyright © 1982 by
ACADEMIC PRESS INC. (LONDON) LTD.

All Rights Reserved

No part of this book may be reproduced in any form,
by photostat, microfilm or any other means,
without written permission from the publishers

British Library Cataloguing in Publication Data

Recent advances in statistics.
 1. Mathematical statistics—Congresses
 I. Oliveira, J. Tiago de   II. Epstein, B.
 519.5   QA276.A1

ISBN 0-12-691580-6

Typeset by Advanced Filmsetters (Glasgow) Ltd
and printed in Great Britain by Thomson Litho Ltd, East Kilbride, Scotland

CONTRIBUTORS

Richard E. Barlow, *Operations Research Center, University of California, Berkeley, California 94720, USA.*

H. A. David, *Statistical Laboratory and Department of Statistics, Iowa State University, Snedecor Hall, Ames, Iowa 50011, USA.*

Bradley Efron, *Department of Statistics, Stanford University, Stanford, California 94305, USA.*

Benjamin Epstein, *Faculty of Industrial Engineering and Management, Technion-Israel Institute of Technology, Haifa 32000, Israel.*

J. Gani, *CSIRO, Division of Mathematics and Statistics, P.O. Box 1965, Canberra City, A.C.T. 2601, Australia (Present address: Department of Statistics, University of Kentucky, Lexington, Kentucky 40506, USA).*

Peter J. Huber, *Department of Statistics, Harvard University, Oxford Street, Cambridge, Massachusetts 02138, U.S.A.*

Dennis V. Lindley, *2 Periton Lane, Minehead, Somerset TA24 8AQ, England.*

Colin L. Mallows, *Statistical Models and Methods Research Department, Bell Laboratories, 600 Mountain Avenue, Murray Hill, New Jersey 07974, USA.*

Emanuel Parzen, *Institute of Statistics, Texas A&M University, College Station, Texas 77843, USA.*

Herbert Solomon, *Department of Statistics, Stanford University, Stanford, California 94305, USA.*

J. Tiago de Oliveira, *Academia das Ciências de Lisboa, Rua da Academia das Ciências 19, 1200 Lisboa, Portugal.*

John Tukey, *Information Sciences Division, Bell Laboratories, 600 Mountain Avenue, Murray Hill, New Jersey 07974, USA, and, Department of Statistics, Princeton University, Fine Hall, Princeton, New Jersey 08540, USA.*

Alexander S. Wu, *Operations Research Center, University of California, Berkeley, California 94720, USA.*

PUBLISHER'S NOTE

This work has been put together as part of the 200th anniversary celebrations of the Portugese Academy of Sciences. A special limited edition is to be published by the Academy in their series "Frontiers of Knowledge". We are grateful to the Academy for their agreement to the publication of this general edition and congratulate them on their anniversary.

PREFACE

Founded on 25 December 1779 as the Academia Real das Sciencias de Lisboa, the Academia das Ciencias de Lisboa decided to celebrate its second centenary by a series of activities, from the publication of the "State Papers" and the colourful "Livro das Armadas" to the sponsorship of a series of symposia "Frontiers of Science".

The symposium, "Some Recent Advances in Statistics" was one of this series. We use the word "some" in the title because it was not possible in a subject as broad as statistics to include all recent advances, within the time and resources allotted to the symposium. Nevertheless, the topics presented at the symposium covered a spectrum sufficiently wide to demonstrate again and again how probabilistic and statistical thinking and methodology influences and is influenced by virtually all fields of human endeavour. Among the topics covered are: stochastic epidemic theory, statistics and the law, statistical data modeling, exploratory data analysis, computer based statistical methods, robust statistics, Bayesian view of statistics, concomitants of order statistics, extreme values, life test sampling plans, analysis of incomplete life length data.

Amidst the diversity of the contributions to this book, there is also an underlying unity. We hope that it represents another step toward the integration of various trends in statistical thinking and that it will serve as a bridge linking statisticians with the various disciplines represented in the Academy.

J.T.O.
B.E.

CONTENTS

Preface	vii
1. Measurement and Burden of Evidence. HERBERT SOLOMON	1
2. Data Modeling Using Quantile and Density-Quantile Functions. EMANUEL PARZEN	23
3. Stochastic Processes in the Theory of Epidemics. J. GANI	53
4. The Bayesian Approach to Statistics DENNIS V. LINDLEY	65
5. Concomitants of Order Statistics: Theory and Applications. H. A. DAVID	89
6. Decision and Modelling for Extremes. J. TIAGO DE OLIVEIRA	101
7. An Overview of Techniques of Data Analysis, Emphasizing its Exploratory Aspects COLIN L. MALLOWS and JOHN W. TUKEY	111
8. Computer Intensive Methods in Statistics BRADLEY EFRON	173
9. Current Issues in Robust Statistics PETER J. HUBER	183
10. Bayesian Evaluation of Life Test Sampling Plans. RICHARD E. BARLOW and ALEXANDER S. WU	197
11. The Statistical Analysis of Incomplete Life Length Data BENJAMIN EPSTEIN	213

1
MEASUREMENT AND BURDEN OF EVIDENCE

Herbert Solomon

Abstract

In this essay, issues arising in administrative, civil and criminal courtroom settings will be featured with regard to magnitude of allowable risk and also with regard to the employment of rather sophisticated concepts in statistics and probability that in some cases are rather central to a determination by the court. The two issues are of course not independent, but the admissibility of statistical and probabilistic thinking in courtroom settings, while not new, is catching on and will probably burgeon in the near future.

The term "beyond a reasonable doubt" has always puzzled jurors and judges who sit in criminal courts. The question of how to evaluate the concept or even provide a numerical index has always seized the imagination. This is also true for other burden standards such as preponderance of evidence, clear and convincing evidence and clear unequivocal and convincing evidence. An account is given of a recent case in the US Federal Court System in which an attempt is made by a judge through a survey of his colleagues to relate various degrees of burden of proof to exact probability figures.

Trials should not be accomplished by mathematics and statistics, but it is hopefully shown that there are special situations where statistical thinking is central to an issue, and that in these cases it is foolish, and sometimes even unfair, not to use it.

1. Introduction

Measurement of evidence and the burden this evidence should bear in public policy decisions receive much wider attention then ever before. These require the quantification of risks to individuals and society which in some cases may be easily achieved but difficult to employ in decision making and in many situations both measurement and employment defy the thrust of the investigator. For this exposition we will look into measurement and quantification of risk in civil law and criminal law settings, but analyses in the broader context must necessarily seep into the discussion. In the BBC 1978 Richard Dimbleby Lecture, Lord Rothschild offered "Risk" as his title and essentially discussed how slight a risk would be before an individual or society would ignore it. Of course, there is no textbook answer, and the best that one can do, if possible, is to provide prior situations where probability values are either achieved through scientific reasoning, subjective estimates or pure guess and decisions have been made. For example, is the convenience of riding in automobiles worth a one-in-five-thousand chance of being killed in any given year? If so, does it follow that a risk of only one in 750 000 is worth taking in exchange for the benefits of a nuclear power plant? As happened in the 1979 summer, there is a small risk that an orbiting satellite will kill someone when it falls to earth. Does this justify the immense cost of altering its path and controlling the point where it will come to rest? Obviously the weight to be given to a risk value is inextricably tied to the context in which it is viewed.

Conventional wisdom suggests that zero risk is best. However, to demand this would give us a quite different world and one that we would probably not tolerate. The technology we now possess and new developments would be doomed. Political and social upheavals would follow the refusal of society to take risks. Given that zero risk is impossible or at least imprudent, great questions and complex issues arise in arriving at decisions on amount of risk to be tolerated by the individual and society in such matters as amount of carcinogens ingested in food versus their ability to improve food production, impact of side effects of drugs versus their ability to cure disease or aid in nutrition or in controlling population growth. What always comes to the fore in such matters, as we will see in civil and criminal settings, is a value for the acceptable level of risk. For example, the chance of being killed in an auto-accident in Great Britain is about 1 in 7500 per year; yet, this does not seem to deter the population from motoring about the country. Does this suggest if the risk of being killed was greater than (say twice) this value, that the risk would be considered too great?

As in other matters, while this subject is receiving much attention these days, it is not a new consideration. The French savants of the late eighteenth and early nineteenth centuries gave serious attention to the magnitude of a risk that is allowable or which one may neglect with propriety. Condorcet (1743–1794) realized that the magnitude of a risk to be neglected in any situation varies with the subject matter to which it relates. This kind of thinking offers problems in Anglo-Saxon jurisprudence especially as it is practiced in the United States where judge or jury is not supposed to consider penalty in reviewing the burden of evidence.

2. Risk measurement

Condorcet tries to look at risk in several categories; for example, (1) the impact of an incorrect decision between claimants as to the right to a property, and (2) the condemnation of an accused person to capital punishment. On the latter category, we may observe that Condorcet is of the opinion that capital punishment ought to be abolished on the ground that however large may be the probability of the correctness of the single decision, we cannot escape the large probability that in the course of many decisions, some innocent persons will be condemned. In short, in the case of the death penalty, Condorcet wishes to eliminate any impact of the variability that must exist in the guilt/innocence decision of judge or jury from true guilt and true innocence. Others throughout the ages, e.g. Maimonides, have essentially said, "It is better to permit the crime of a guilty person to go unpunished than to condemn one who is innocent." In 1979, the British Parliament voted that the death penalty not be considered in a criminal trial and thus reaffirmed a similar decision made earlier. However, in the United States, capital punishment is once again permitted in many of the states by Supreme Court decisions.

For the category on property rights, Condorcet notes an opinion expressed by Buffon (1707–1788). Buffon states in his work that out of 10 000 persons one will die in the course of a day. But for all practical purposes, the choice of dying in the course of a day is disregarded by mankind so that 1 divided by 10 000 may be considered the numerical estimate of a risk which any person is willing to neglect. From the actuarial tables of Buffon's day, this is the chance that a man 56 years of age will die in the course of a day. Note that for everyday motoring, we accept 1/5000 or 1/7500 as risk values we are willing to neglect in order to enjoy that technological innovation. However, I doubt that in the USA, the Food and Drug Administration would accept 1 out of 10 000 as a tolerable risk if it applied to a bladder cancer rate from say the use of saccharin in our diets.

To return to Condorcet, we find that he objects to Buffon's estimate and proposes the following: from mortality tables of his day, he finds the risk of sudden death in the course of a week for a person aged 37 is 1 over 52 times 580, and for a person aged 47, is 1 over 52 times 480. He assumes that practically no person distinguishes between these risks so that their difference is in fact disregarded. This value of approximately 1 divided by 150 000 (exactly 1/144 768) is proposed, therefore, as a risk which a man would consider equivalent to zero in anything affecting his own life. In terms of insurance rates for death in aircraft accidents at present, the risk ratio is closer to Condorcet's value than that proposed by Buffon. Note, however, that even if the probability of innocence of a murder is that small value, Condorcet will still not permit execution.

What is of central importance here is not the exact value associated with the risk but the fact that the French mathematicians of those days were so acutely aware of the problem. The French school of probabilists were caught up in the swirl of studies on social behavior, especially jury size and jury decision making, from the time of the French Revolution until the mid-nineteenth century. Poisson's book published in 1837 was on the calculus of probabilities but featured this kind of thinking in civil and criminal proceedings. It is in that book that the Poisson distribution first appears. Cournot, Bienaymé and Laplace also contributed in this period (we have already mentioned Buffon and Condorcet who preceded them), although Laplace's efforts in this subject were minor. This group was criticized by several savants in the nineteenth century for attempting to employ the calculus of probability in human behavior models. It may be instructive to provide here a quote from a twentieth century scholar, R. A. Fisher, who exhibited similar sentiments about probability theory in our time by suggesting that we are still tied to that period.

> This mid-twentieth century is not the first period of grave confusion in the teaching of the Theory of Probability. For fifty years from the publication of Laplace's *Theorie analytique*, books in France and England were full of rubbish concerning the veracity of witnesses, and the probability of correctness of the findings of tribunals. The present confusion seems largely to be a hangover from that period from which nineteenth-century discussion had largely extricated mathematical thought in France and England, but perhaps less completely in some more distant countries.
>
> There is the difference, however, that whereas in the nineteenth century error could be rife in mathematical departments, without doing greater harm than to confuse their students, in our time really important matters such as the standardization of drugs, the control of epidemics, and the precision of ballistic missiles are liable in the future to be influenced by young men now leaving these departments armed with erroneous numerical tables, as well as with confused and obsolete ideas. This, in some sort, concerns us all.

I would tend to look at the period from mid-nineteenth to mid-twentieth century as barren in probabilistic thinking for human behavior settings except for possibly the British school of psychologists who developed factor analysis models in the late nineteenth and early twentieth centuries as a tool in the measurement of human intelligence. However, there is a resurgence in several countries such as the United States, the United Kingdom, France, Scandinavia and Israel prompted by societal issues. Those investigators who toil in the behavioral sciences in these times still receive the scorn or apathy of a number of probabilists and this is compounded by disdain from social scientists who resent these interlopers. These constraints plus the intractability of a number of problems limit the number of workers in this field.

Before I get into some details of risk measurement in the civil and criminal setting, let me point out a decision in a recent case, *US* v. *Lopez* USDC ENY 328 F.Supp 1077 (1971), in the United States which talks to this issue. Two men were arrested at Kennedy Airport in New York City on the charge of possessing narcotics. At issue before the Federal Court was whether the search that produced the narcotics was legal. Before boarding an aircraft, the two were asked to step aside by the security people for Pan Am, at which they bolted and ran. They were pursued, caught and searched. The reason given in court for their being asked to step aside was that additional scrutiny was required because of a profile analysis employed on boarding passengers by Pan Am. The exact variables employed in the profile were revealed only to the court *in camera*. The judge ruled that use of a profile analysis singled out some passengers from others, but was not unconstitutional because from previous data supplied by the Government it appeared that 6% of all passengers flying in the United States who were asked to step aside to be searched carried firearms or other dangerous weapons, and that public policy therefore required such action. It was also asserted that 1/10 of 1% of all passengers were asked to step aside for scrutiny. Note here that a value of 0.006 of 1% of all plane passengers is a magnitude of risk that the court is not willing to neglect and that obviously it is tied to the seriousness of the situation wherein hijacking or related offenses may ensue. It should be obvious that in another context the court could rule that this is a magnitude that could easily be neglected. In fact the court so stated that if 6% of the general population carried firearms, police could not arbitrarily ask an individual walking along a street to step aside for search under the US Constitution. I should add that in this particular case the two defendants were not convicted because the court learned that the profile contained variables directly related to ethnic and racial considerations, and in this sense, was unconstitutional; and that therefore the evidence obtained by the search was illegal and could not be used in court. However, the profile idea is still under way in the United States and in many other countries, since the seriousness of a hijacking event far outweighs the inconvenience to the public, despite the small probability of a hijacker appearing in the population of plane passengers. I would be remiss if I did not mention at this point that the profile itself was constructed by discriminant analysis—a technique developed by R. A. Fisher.

In the remainder of this essay, we will feature issues arising in administrative, civil and criminal courtroom settings with regard to magnitude of allowable risk and also with regard to the employment of rather sophisticated concepts in statistics and probability that in some cases are rather central to a determination by the court. The two issues are of course not independent, but the admissibility of statistical and probabilistic thinking in courtroom settings, while not new, is catching on and will probably burgeon in the near future.

3. Criminal case illustration

Let me begin with a case, in which I served as an expert for the defense, that is not very dramatic but which highlights points I wish to make. In the State of California, gambling for value in money or in kind is banned for some games (e.g. roulette), is legal for other games (i.e. draw poker) and, where games are not mentioned in the Act, the issue is whether skill in a game predominates. The game of pinball is one that is not mentioned by name in the law, and a number of years ago, several shopkeepers were arrested in San Francisco on the charge that they had paid a prize in money to players (San Francisco policemen in mufti) on pinball machines. The State considered pinball a game of chance, that is, skill did not predominate in achieving a high score (a win) in the game.

For any game not mentioned in the law, the issue as to whether skill predominates depends, of course, on the particular game. Thus the logic by which one can show that skill either predominates or does not also varies with each game. In pinball, one or more of five metal balls are sent on their way by a plunger until all five have been dispatched, and the player hopes to achieve their falling into holes which give high scores, so that a total score achieved for all five can pass a criterion point indicating a win. The distance the plunger is pulled back and the manner in which it is released, plus the application of hand pressure to the body of the machine, influence the course of the metal balls and thus the score. Too much pressure causes the game to terminate abruptly, and this is recorded as a loss.

To see whether skill predominates, the following experiment was developed. A number of games was to be played by a robot (graduate student) to determine the innate probability of a win built into the machine. An "expert" player would then play the same number of games, and the proportion of wins would be matched against the proportion of wins achieved by the robot. One question of a strict statistical nature is the number of games to be played. As in all sample size problems, one desires a size which is not too large to be costly in some sense, and, of course, not too small, for then the variability in the results could overwhelm any true differences that might exist. Statistical analysis suggested that 1000 games played by the robot and by the expert would suffice in that, for this binomial situation, the variance would be quite small (at most, 1/4000). Moreover, 1000 games was feasible for the robot (graduate student) and the expert. The ratios of winning performance by the robot over three similar but different pinball devices ranged from 0.05 to 0.10 and correspondingly, the winning performance ratios of the expert ranged respectively from 0.10 to 0.20. Since the variability is quite small, one can assume for all practical purposes that these are the values, and the question is, how to make a determination of skill. It should be mentioned at this point that conviction for this criminal misdemeanor in the early 1960s carried with it a jail sentence up to 6 months, or a maximum fine of $500, or both.

We are now back to the issue considered earlier of when is a difference small enough to be disregarded in terms of risk to the individual or society? In this particular situation, we have differences between the robot and the expert running from 0.05 to 0.10, or equivalently one can say that the expert does twice as well as the robot. The evidence is now measured, but how does one view the burden of the risk that is now measured?

As an expert witness, my reasoning led me to state that the expert had demonstrated that skill predominates. As an example, I brought before the court the decision made in the United States in the mid-1950s that individuals in the population should receive the Salk polio immunization vaccination. There was something very similar in the logic of that experiment and that for the pinball case. In the Salk clinical trials, young children were given immunizations which were double-blind, that is, neither the doctor nor the parent or child knew which child was receiving a placebo or the dead virus. For those receiving the placebo one can imagine nature playing its game with this deadly disease, and this resulted in an incidence of 40 per 100 000 cases of paralytic polio (0.000 40). The expert playing the game (Dr. Salk) achieved an incidence of 15 per 100 000 (0.000 15) for those children who received the dead virus. The difference here is only 25 per 100 000 (0.000 25) and yet the Surgeon-General of the United States asked that everyone first under the age of 40 and then all, receive the immunization. It is obvious that if these differences occurred in testing a common cold immunization or for testing a drug for some minor skin ailment that disappeared in 2 or 3 weeks anyway, such a proclamation would not have been advanced by the Surgeon-General. It seemed to me, therefore, that the differences determined in the pinball experiment, when associated with loss of liberty (county jail) or loss of property ($500 fine), were large enough to suggest that skill predominated. If the maximum penalty were a $10 fine, I doubt that anyone would care one way or another, despite the evidence that had been amassed and analyzed. Yet to consider the penalty as part of the consideration in assessing the evidence renders invalid the statement of the expert witness in the criminal setting, at least in a number of US courts.

Before we go into what might be more essentially probabilistic situations, let me offer another example of the use of statistical thinking, in this case, by the US Government to demonstrate evasion of payment of rather large income taxes, and therefore leading to a claim for large sums from the taxpayer. Just as in the pinball case, the evidence issues here will be essentially statistical. In other words, we are shying away from cases where statistical thinking may play some minor role and we are trying to feature those situations where statistical or probabilistic thinking is central to the legal situation.

4. The confidence interval approach in civil matters

The Internal Revenue Service reconstructed the cash drop at crap tables at a Las Vegas casino for 3 fiscal years beginning April 1, 1961, and ending March 31, 1964. The basis for the reconstruction is the cash count obtained by teams of IRS agents at each of 87 crap table shifts at sampling intervals over this period of time. On the basis of these counts, the IRS alleged that shortages in reported income occurred according to some design by the casino. On the basis of this sampling of a population of over 19 000 crap table shifts for the 3-year period, the IRS asked for several million dollars in delinquent income taxes, plus a 50% penalty that alleged fraud. Despite the pejorative connotation of that term, the action was a civil procedure rather than a criminal action that had been dispensed with earlier for other reasons.

It appears upon examination that the IRS was a bit high-handed in the way it handled the statistical assessment of unreported income. It claimed at first that a team of three agents spelled each other over an 8-h shift and counted the cash presented to the table for either gaming or for chips. All cash is immediately put into a drop box at the table and at that time was then counted after the 8-h shift by the casino and posted in their ledgers. The totals provided by each of the three agents were then accumulated and matched with the house count. The comparison was done at a much later date—for one thing, an income tax return is filed only once a year and the IRS auditing occurs afterwards. Since the agents were informed to count all bills as $1 bills if they could not read the denomination, the assumption by the IRS was that, at best, their count should be under what the house reported. When comparisons were made over the 87 shifts, the house count was, at times, under the amounts recorded by the IRS agents. This suggested under reporting of income received ("skimming") to the IRS.

When the issues began to receive scrutiny by both sides, and depositions were taken, the IRS stated that two teams of three IRS agents made the counts rather than one, and that, moreover, that team reporting the larger income provided the value used by the IRS. The IRS gave no thought to individual variability in counts made under casino gambling conditions. This explicit lack of attention to counting errors made by IRS agents is a fundamental mistake in the basis for their claims. It stands out sharply because the IRS itself assumes the possibility of such counting errors by requiring the use of two teams of agents to estimate cash transactions at each crap table and an egregious error in the IRS procedure is the use of the maximum of the two cash counts arrived at by the IRS teams as the actual casino income. The use of the maximum value in this situation is an especially poor decision as students in elementary statistics classes are usually taught. It can be demonstrated by a very simple model that the employment of this rule can lead to a probability as high as 0.75 that the casino can be accused of understating its income, when in actuality it is not.

The central issue here, however, and it is a statistical one, is how to measure the influence of the counting error on estimating the true income. It was possible, from all the data available, to arrive at an estimate of the standard deviation of counting error made by the IRS agents. This turned out to be approximately $690. A 95% confidence interval or 99% confidence interval would give us two or three times this amount divided by the square root of two on either side of the sample mean (based on a sample of size two) as a range in which, with the stated confidence, the true value (actual cash drop per 8-h shift) could fall. That is, swings of $840 or $1260 on each side of the mean value obtained from the two counts of the IRS teams. When this confidence interval procedure is applied over the 87 shifts, we find that with a lower one-sided 95% confidence interval, 75 times out of 87, the house count fell inside the interval, and for a lower one-sided 99% confidence interval, 82 out of 87 values fall inside the interval. This is also accomplished for two-sided 95% confidence intervals and we find that 17 out of the 87 shifts fall outside the interval, and for two-sided 99% confidence intervals, we find that 9 out of the 87 shifts fall outside the interval: in the first case, 9 cases in which the house count is too large and 8 cases in which the house count is too small, and for the second situation, 4 cases in which the house count is too large and 5 cases in which the house count is too small.

Arguments can be presented for either one-sided or two-sided confidence intervals in this analysis. The two-sided confidence interval makes sense if one is looking for deviations from house count in either direction—namely, human counting errors are just as likely in one direction as another around a true value. The one-sided confidence interval could be based on the mechanism in the IRS instructions by which some cash drop may be missed altogether, or reported as a lower value whenever there is doubt about actual magnitude. Under these alternatives, the true cash drop is likely to be larger than IRS reconstructed cash drop, provided there is no IRS fallibility in counting.

One could go on with more statistical tests that help resolve the situation. What is important, from the point of view of measurement of evidence, is that statistical analysis suggests that differences between IRS counts and house reported income fall well within the yardstick of inherent variability in human counting errors experienced by IRS agents under casino gambling conditions. This does not mean that the casino was reporting income correctly. It demonstrates only that the evidence provided does not eliminate human counting error as the reason for the differences. Also it does not prove or disprove concerted "skimming" by the casino.

The case offered provides evidence based wholly on statistical analysis, a method central to the Government's approach, which when applied correctly could have suggested to the Government that its basis for winning the claim did not have a good evidentiary foundation. As is usual in these cases, the motivation for the proceedings contained factors that might be considered to be somewhat political in nature, since it resulted from the drive of the Attorney-General in the early 1960s to get at organized crime in Las Vegas through the IRS.

The use of confidence intervals in legal settings has appeared in other contexts. In an article by Katz (1975), another instance is given. In that article Katz discusses his analysis of a very specific case in the State of Michigan in which he was an expert witness, and also some of the problems faced by a statistician in a courtroom. The issue at hand was to provide an estimate of the number of uninsured automobiles in the population of passenger cars. It was illegal to drive a car without insurance in Michigan and so this would obviously complicate the response to be received from the owner of a car that fell into the sample. These individuals had to be assured that the survey was for informational purposes only and that no action would be taken against them if they responded. Katz traces how in the sample of 249 vehicles out of a total population of over 4 000 000 he arrives at a confidence interval estimate of roughly 2–6%. How to present this analysis and its results are the interesting part of the paper since in effect the statistical analysis of the data obtained by systematic survey sampling is essentially a textbook operation.

Still another example of survey sampling and its admissibility in a court of law is given by Sprowls (1957). In that particular situation an expert witness presented a confidence interval on the proportion of out-of-city customer sales at a Sears-Roebuck store in Inglewood, California. At issue was the amount of sales tax to be paid since it was not due except for purchases by residents. A 95% confidence interval for the ratio of out-of-city sales to total sales was given with endpoints of 33.7 and 39.7%. This translated into a dollar value of about $24 000–$32 400 for the tax amount at issue. The judge ruled against the introduction of these sampling results as evidence but said he would permit a complete audit before he ruled on the case if the store was willing to undergo this expense. The complete audit yielded a figure of $26 750.00 which was just a bit short of the original claim of $27 000.00 and of course lies nicely in the 95 % confidence interval. In this situation the court was in effect giving zero weight to a random variable, that is the sample estimate of 36.69 % as a piece of evidence. This is an atypical situation because it is one where the population can be reconstituted and the resources are on hand to accomplish this. Usually one or both of these factors are missing. For example in the dice situation reported previously it was not possible to reconstitute the total population of 8-h shifts at the dice table.

Another situation like this in which I had some direct experience concerned the reading of electricity demand meters. A large sum of money hinged on what the actual readings were for some electricity demand meters on a specific date. The contestants were the Pacific Gas and Electric Company and the Sacramento Municipal Utility District, two large agencies. What was at issue was the actual reading on each of four meters at a certain point in time and of course the actual demand at that time could in no way be reconstituted. It was possible from engineering data to derive a confidence interval on what the true demand was at that time but the question of admissibility of this kind of evidence never came to the fore in this case because the matter was settled by the two litigants without recourse to a judicial decision. I would like to believe that the report I prepared had some influence on the settlement before a judicial decision. A more detailed discussion of this case, the previously discussed pinball case and some other situations of statistics in the courtroom, can be found in Solomon (1966).

5. Multivariate data analysis in a criminal matter

A rather sophisticated use of statistics in a criminal court setting was reported by Niblett and Boreham (1976). The central issue here was the trustworthiness of certain oral confessions allegedly made by the defendant. Naturally, the defendant had denied making these statements to the police and they were extremely important to the prosecution. Not many cases will appear in which it would even be possible to use the technique we are going to describe, yet it was possible to accomplish in this particular case. Once again, we hark to our theme. We do not wish trials to be accomplished by mathematics and statistics, but rather that, in those special situations where statistical thinking is central to an issue, it is foolish, and sometimes even criminal, not to use it.

The defendant was alleged to have made certain oral statements that were recorded verbatim by the police interrogator. The confessions in the statements, all of which were disputed by the defendant, were crucial to the prosecution case. It so happened that the defendant had previously been prosecuted for murder, of which he was acquitted, and on that occasion had made statements to the police, the authenticity of which was not in dispute. Here we have the advantage of an anchor against which we can evaluate the oral statements in contention. For the two sets of statements, namely, those at the murder trial and the statements allegedly made in connection with the present case, were made in very similar circumstances, that is, in a police station under interrogation and recorded verbatim. The defendant felt that computer analysis of the two sets of statements might be able to test his assertion that the second set was not his. Accordingly, the defense called on an international authority on stylistic analysis to give expert evidence, the Rev. A. Q. Morton, and he, in turn, called on B. Niblett to present evidence of the analysis by computer of the two sets of statements.

Previously, Niblett and Boreham had completed the development of a set of computer programs for classifying the text of lengthy documents, for example, the conventions and agreements of the Council of Europe—some 280 000 words of text in both the English and French versions. To accomplish their classification, the full text of a document is converted into digital form so that they can be read and processed by a computer. From each document, the computer is programmed to construct a document vector, that is, a string of numbers that corresponds to the words in the document and the frequency with which they occur. Once these vectors of all documents in the set are constructed, each vector can then be compared with all the others by forming a measure of association or, in another parlance, the scalar product of the vectors, to give a measure of similarity of the documents based on the words they contain. These coefficients of similarity all lie within the range 0–1, a coefficient of 1 means that the documents have identical content, whereas a coefficient of 0 means that the documents have no words in common. The values in between indicate various degrees of overlap of documents. From coefficients of similarity, a similarity matrix can be constructed, and this is usually the basis for clustering procedures. In this particular case, of course, the issue is whether the document vectors that result from the conversations during the interrogation before the murder trial cluster, are disparate from the document vectors that arise out of the interrogations submitted by the police at the present trial. A complete delineation of the sets of vectors resulting from the second trial from those of the first trial would certainly be in the defendant's favor.

One problem, of course, is the length of the document vector. To make it too long would make it impossible for computers to do the job in connection with clustering algorithms. Since the number of coordinates in the document vectors has to be restricted, questions arise as to which words or what kinds of words should be included and, of course, in this context, the question is, what are the words that best discriminate? Previous studies, such as those by Mosteller and Wallace in their investigation of the disputed authorship of some of the Federalist papers, have used function words as discriminators of style. Function words are in contrast to context words. They are usually articles, conjunctions, prepositions, etc. For this particular situation, the computer was programmed to include only the 80 most common function words, and these were selected on the basis of previous experience by Niblett and Boreham.

The 11 statements provided by the defendant were then analyzed. Documents 1–7 (the undisputed documents) contained the statements in his own words that the defendant agreed he made under police interrogation in the murder investigation. Documents 8–11 (the disputed documents) were statements the defendant was alleged to have made, but which he denied. For the purpose of the analysis, the identification of each of these documents is lost, and so 11 documents are taken together as one set and converted by the computer into 80-dimensional document vectors. Similarity coefficients are then calculated for the 55 different pairs of documents $(11 \times 10)/2$. These are then subjected to a cluster analysis. The particular clustering procedure adopted was an agglomerative hierarchical method using the nearest neighbor or single linkage criterion. Because of the hierarchical nature of the procedure, each document can appear in only one cluster. The results of the clustering are then exhibited in what is known as a dendrogram which shows the clusters formed at any given level of similarity coefficient. Ultimately, all the documents will merge into one global cluster. Viewing the resulting dendogram (p. 178) documents 8, 9, 10 and 11 cluster together in one group, documents 5, 6 and 7 in another group, 3 and 4 in still another, and then 1 and 2 as isolates. However, 1 and 2 join 5, 6 and 7 subsequently, and 3 and 4 join these five, so that the final picture shows two quite disjoint clusterings, one containing the undisputed documents (1–7), and the other the disputed documents (8–11).

A question now arises as to whether this result could be spurious. To answer this, Niblett and Boreham used exactly the same technique on documents of similar length whose authorship was known, in order to test the reliability of the method. Two separate situations were investigated. In one case, six sonnets by Shakespeare and six by Wordsworth were analyzed, and in the other case, there were six collects from the *Common Prayer Book* of the Church of England and six extracts from the Damon Runyon novel *Guys and Dolls*. The dendrogram for the Shakespeare and Wordsworth sonnets showed a bit of overlap, so no well-defined clusters discriminating between Shakespeare and Wordsworth were found. Certainly, then, the cluster technique is unable to distinguish the two poets, based on the evidence provided by their sonnets. On the other hand, the dendrogram shows quite separate clusterings for the collects from the *Common Prayer Book* (Church of England) and the extracts from *Guys and Dolls*. As expert witnesses in the court, Niblett and Boreham stated there was a distinct and substantial difference in style between the disputed and undisputed documents, and that the difference in style, based on the occurrence and frequency of function words, was greater than that between Shakespeare's and Wordsworth's sonnets, but less than that between the *Common Prayer Book* and Damon Runyon. Thus, if function words are accepted as an indicator of style, the evidence here is that the disputed and undisputed documents are unlikely to be by the same author. In this case, the defendant was acquitted on eight of the twelve counts brought against him, and it was in these eight that the alleged confession statements were relevant.

6. Probabilistic thinking in criminal matters

We now look at two criminal cases in which probabilistic thinking is central to the decision made by the judge or jury. The first occurred in the State of California in the late 1960s and now for over 10 years has received a lot of attention through a spate of articles in law journals and journals dealing with social psychology, political science and statistics. It is known as the Collins case. Mr. and Mrs. Collins were tried for second degree robbery in Los Angeles and were found guilty by the jury hearing the case. Evidence was presented in the case that purse-snatching had occurred by a couple who made their getaway in a yellow car and who allegedly answered the description that fitted Mr. and Mrs. Collins, namely, a bearded black man with a mustache and a Caucasian girl with blond hair and a pony tail.

It was demonstrated by an expert witness for the prosecution (no evidence other than this was essentially prepared for the jury) that the chances of a couple such as the one appearing before the jury were 1/12 000 000 and this was accomplished in the following way. The prosecuting attorney, though he insisted it was only for illustrative purposes, asked the expert witness how he would compute the probability of finding a couple jointly possessing the following characteristics under the following situations: (a) partly yellow automobile—individual probability 1/10, (b) man with mustache—individual probability 1/4, (c) girl with a pony tail—individual probability 1/10, (d) girl with blond hair—individual probability 1/3, (e) black man with beard—individual probability 1/10 and (f) interracial couple in car—individual probability 1/1000. The expert witness stated that one just multiplies these probabilities together and therefore there could only be one chance in 12 000 000 that such a couple exists.

Both defendants had taken the stand on their own behalf and denied any knowledge of participation and also offered alibis which could not place them near the scene of the robbery when it occurred. Upon appeal, the California Supreme Court reversed the conviction, suggested it could do so on several grounds, but featured one of obvious interest to us in connection with this exposition, see *People v. Malcolm R. Collins*, Supreme Court of California in Bank, 68 C.2d 319, March 11, 1968.

First, from a statistical point of view, the estimates provided by the prosecutor were not based on any demographic studies. It had been previously held in other courts (e.g. the Sneed case in New Mexico) that mathematical odds are not admissible as evidence to identify a defendant where the odds are based on estimates, the validity of which had not been demonstrated. The Supreme Court also went into the question of the independence of the characteristics selected by asserting that no proof was presented that these characteristics were mutually independent, which would be required if all six probabilities can be multiplied together to give the probability of the joint occurrence. In fact, in this case, it can easily be seen that some traits or characteristics are not mutually independent. For example, black men with beards and men with mustaches obviously represent overlapping categories. In simple language two defects already exist: (1) inadequate evidentiary foundation of the odds and (2) an inadequate proof of statistical independence.

We now, along with the California Supreme Court, turn to the most fundamental error, that is, the one caused by the probability testimony. A quote from the Court's decision would be instructive at this point:

> We think that the entire enterprise upon which the prosecution embarked and which was directed to the objective of measuring the likelihood of a random couple possessing the characteristics allegedly distinguishing the robbers was greatly misguided. At best, it might yield an estimate as to how infrequently bearded negroes drive yellow cars in the company of blond females with pony tails.

The crucial issue, according to the Court, was "of the admittedly few such couples, which one, if any, was guilty of committing this robbery?" The Court mentions that there are a number of risks of error permeating the prosecution's circumstantial case. For example, the guilty couple might have included a light-skinned negress with bleached hair rather than a Caucasian blond. Or the driver of the car might have been wearing a false beard as a disguise. Or the prosecution's witnesses might simply have been unreliable. Despite all this, and even accepting the final statement about the probability of 1 in 12 000 000 as being arithmetically accurate, the Court could not see how one could still conclude that the Collinses were probably the guilty couple using the reasonable doubt criterion.

It is to this latter point that we now look at probabilistic thinking as central to the Supreme Court reversal. It is in this connection that the support goes to the basic index of proving guilt beyond a reasonable doubt. In trying to assess the magnitude of doubt in this case, the Court set itself the following exercise. What is the probability that this set of joint characteristics will occur more than once in a group of N couples in which the joint characteristics occur at least once? To evaluate this probability, the Court is willing to go along with the prosecution and assume that the unconditional probability of a couple with these characteristics appearing in the Los Angeles area is 1/12 000 000. Assuming that N is large, as it would be in the Los Angeles area, the Court, in a technical appendix to its decision, evaluates this probability as $(e-2)/(e-1)$, which is approximately 40%. Thus, since there is a 40% chance that another such couple exists, the reasonable doubt criterion is satisfied, and the conviction was reversed. In a subsequent section we look into quantifications of the reasonable doubt index.

A number of investigators have chastised the California Supreme Court for this analysis. Usually a Bayesian approach is taken. This alludes to some distribution of the proportion of couples in the Los Angeles area answering to these joint characteristics. Obviously if the true mean value is 1 in 12 000 000 (and of course this was never really shown) the variability around this value could be exceedingly small, and thus it may well be that the 40% value may be somewhat off the mark. I do not believe it was the Court's intention to worry about different prior distribution models that might be applied to the case. What troubled them obviously was not the use of statistics and probability, in fact they make this statement, but the fact that it was employed in a logically irrelevant and evidentiarily inadequate manner. Moreover, it was a criminal case in which essentially no other evidence was applied. It was probably one of those situations which never should have been brought into the criminal courts. It was probably the shoddy and inept manner by which the prosecution attempted to secure a conviction by the abuse of statistical and probabilistic thinking that offended the Court, especially where this was the only evidence available to the prosecution for conviction. We have not heard the last of this case in the literature.

Before we go into a very recent case on the use of probability to assess evidence in a criminal case, I would like to mention one rather old example and one somewhat similar to the Collins case. The Sneed case in New Mexico in the 1960s bears some resemblance to the Collins case; see *State* v. *Sneed*, 76N.M. 349 [414P.2d 858] (1966). The defendant Sneed was charged with murdering his mother and father. Testimony revealed on the day of the murder an individual listed as George Crosset registered at a motel in the town where the crime took place and the defendant also admitted that he used that alias on occasion. There was very little other evidence to substantiate the charge. The prosecution brought in an expert witness who during examination testified that the christian name George occurred approximately once in thirty times. As for the surname, Crosset, he had made an investigation of the telephone books in the region and had not found it listed but assumed therefore that the chances could not be more than 1 in 1 000 000. The expert witness then testified that the probability was 1 in 30 000 000 (the product of the two probabilities) of locating a George Crosset and since one sat at the bar then it was very likely that he was the culprit. The New Mexico Supreme Court reversed the guilty verdict on the basis that a numerical value had not been properly established for the existence of each of the two names.

The older case alluded to above is the Dreyfus affair. It is interesting that in the late nineteenth century at a court martial in France that the prosecution presented some testimony using probability theory to prove its case of espionage against Capt. Dreyfus. At the original hearing in 1899, Bertillon had testified that he had made a frequency count of the letters of the alphabet as they appeared in the words in the "bordereau"—a collection of documents that served as a basis for the espionage trial. This frequency count was then compared with the modal frequencies of the letters of the alphabet as they appeared in the collection of the French Academy which was based on the prose writings of a number of French authors in the nineteenth century. Bertillon had asserted how atypical the "bordereau" frequencies were when compared with the Academy proceedings and that this indicated that some secret code had been employed. For the rehearing in 1904 the Dreyfus family and friends had the services of a committee of three distinguished mathematicians, Darboux, Appell and Poincaré. Poincaré indicated that Bertillon's probabilistic reasoning was incorrect, for if it were employed to check any of the French authors whose prose had contributed to the Academy proceedings, each one could probably be convicted of espionage for the same reason, for he asserted that it was not atypical and actually natural for the works of one to differ quite a bit from the modal values. See Affaire Dreyfus: La Revision du Procès de Rennes, Tome Troisième, Ligue Francaise Pour La Défense Des Droits de L'Homme et du Citoyen, Paris 1908. Naturally many other items were contributed by the Army in its prosecution yet it is interesting that probabilistic reasoning was employed three-quarters of a century ago in a criminal proceeding. Actually over 100 years ago in the United States, mathematical probability was employed to determine the chance of coincidence of signatures on a will in a forgery case, see Howland Will Case 4 Am.L.Rev. 625 (1869); and 65 years ago in *People* v. *Risley*, 214N.Y.75, 108N.E. 200 (1915) mathematical probability was employed in connection with whether a typewriter document was forged.

We now go to the District Court in Beer Sheva, Israel, for another judicial decision where probabilistic grounds led to the verdict. An account of this appeared in an article by Rubenstein (1979). In his paper Rubenstein reassessed a prosecution model for conviction. The facts of the case are as follows: A man and his wife were at a cemetery for a funeral. They noticed a wallet on the ground and handed it to the accused, who was present on the scene in his official capacity as a policeman. Shortly thereafter, an acquaintance of theirs told them he had lost his wallet. They referred him to the accused, who then handed over an empty wallet to the owner. The owner complained the wallet had contained Israeli £740 and the policeman was placed on trial for stealing under a section of the criminal code. The prosecution brought forward two main witnesses: (a) the wallet's owner, who convinced the court that his wallet had indeed contained Israeli £740 and (b) the couple who had first found the wallet and who claimed that they saw banknotes jutting out from it. At the trial, the accused claimed that when he opened the wallet, it contained Israeli £1 only.

In coming to his decision, the judge stated that the prosecution's witnesses had made a positive impression, and he also added that it was not conceivable to him, that had the couple themselves stolen the money from the wallet, they would have presented themselves before a policeman and given away their identity. The defense claimed that even if the couple did not lie willfully, there nevertheless exists a possibility that first they had made an error in what appeared to them to be banknotes, and further, that perhaps the money was stolen by a pickpocket, who threw away the wallet after removing its contents. The accused was found guilty, sentenced to 3 months in prison, and ordered to make restitution to the wallet's owner, and of course, this ended his 21 years of service in the police force.

We will now use the judge's summation in arriving at the verdict, for we wish to investigate this in some detail.

> I cannot accept this argument because the chance of both these probabilities occurring together, and I emphasize together, is extremely small. Let me elucidate. Let us assume that the chance that the owner's wallet was stolen is 1/10 (I believe that the chance is, in fact, much smaller). And let us assume that the chance that the prosecution's witnesses erred when they said that they saw the money in the wallet is also 1/10. For the purpose of discussing the defense's argument, I am ignoring for the moment the fact that the prosecution witnesses insist unreservedly that they indeed saw the money with their own eyes. Even so, the chance that both possibilities transpired is not $1/10 + 1/10 = 2/10$, that is, 20%, but $1/10 \times 1/10 = 1/100$. That is, 1%.
>
> In other words, the chance that both situations occurred—as the defense claims—that is, that the wallet was stolen from its owner, thrown away, was discovered by the prosecution witnesses, who brought it to the accused, peeked inside, and erringly, thought it contained money—this chance, according to the theory of chance as I have explained it, is so small that it cannot form a basis for reasonable doubt.

Here again we have a situation where probabilistic reasoning is central to the decision of the court, and the question of a probabilistic value for reasonable doubt is entertained. The court's judgment arouses the same questions we had seen emerge before: (1) Where did the judge take the figures which form the basis for his reasoning? (2) What are the arithmetic computations he carried out? (3) Does the probability of 0.01 which he calculated have any absolute significance that enables him to assert that the innocence of the accused is unreasonable? Suppose the probability was equal to 0.25, what would the judge have decided?

Since the judge has so carefully pointed out his reasoning, it is easier to evaluate and criticize his probabilistic argument. This is what Rubenstein does. The event whose probability is to be evaluated is the policeman's guilt or innocence. The judge could have answered this crucial question directly, but chose instead to derive his answer from an analysis of other questions that he felt could lead to an intuitively better evaluation. He demonstrates that the judge has not considered all possible events and accordingly has not been able to arrive at the conditional probability of the policeman's innocence, that is the probability of the event—policeman stole money—relative to the events remaining after the evidence is given. The ratio of the probabilities that the wallet was stolen to the policeman's stealing the money is given equal weight by the judge in his decision. As this ratio increases, the conditional probability of the policeman's innocence increases. For example, if the ratio is 3/1, the chance of the policeman's innocence rises to 0.25 in Rubenstein's model. Just as in the Collins case, appropriate logic calls for a conditional model yet in each case the court missed this point although each was quite content to admit probabilistic thinking in arriving at a decision.

All this has important implications for the use of probabilistic arguments, especially in criminal settings. In the Collins case and in this case, the fault is not with the use of probability but with the use of poor logic in setting up probability models. In both cases, detailed probabilistic justifications were offered by the prosecutions so that it was possible to look into the logic of their models and reveal poor use of such thinking. If a judge or jury wishes to use this kind of thinking, it is important that it be done in some logical manner, or be subject to reversal on appeal. As in the Collins case where, obviously, it was not proven that they were not the guilty couple, so in this case it is not possible to answer the question whether the policeman indeed stole the money. There is nothing in the evidence which makes this event impossible. However, the issue always is whether reasonable doubt exists. The California Supreme Court suggested that this was so in the Collins case, and in this particular case in Israel, the judge's considerations which he was careful to specify, were quite a bit off the mark, and if he had followed through his logic carefully, he might very well have reached a different conclusion.

7. Reasonable doubt measurement

The term "beyond a reasonable doubt" has always puzzled jurors and judges who sit in criminal courts. The question of how to evaluate that concept or even provide a numerical index has always seized the imagination. I would like to report on a recent case in the Federal Court system in the United States (*United States* v. *Fatico*, USDC ENY 458F. Supp 388 (1978)), in which a court tries to relate various degrees of burden of proof to an exact probability figure. This may have happened before and been recorded but I am not aware of any such happening. The case derives from some proceedings in which the key question is "What burden of proof must the Government meet in establishing a critical fact not proved at a criminal trial that may substantially enhance the sentence to be imposed upon a defendant?" The critical factual issue is whether the defendant, Carmine Fatico, was a "made" member of an organized crime family. In the Fatico case, the defendant, like all other defendants convicted of a crime but not yet sentenced, has a strong liberty interest at stake. Should the court determine that he is a member of organized crime, he will receive a substantially longer prison term. Several criteria which we do not elucidate here are offered by which the sentencing judge comes to a decision.

The procedures by which the facts of the case are determined assume an importance fully as great as the validity of the substantive rule of law to be applied. And the more important the rights at stake, the more important must be the procedural safeguards surrounding those rights. The question of what degree of proof is required is the kind of question which is traditionally left to the judiciary to resolve. Broadly stated, the standard of proof reflects the risk of winning or losing the given adversary proceeding or, stated differently, the certainty with which the party bearing the burden of proof must convince the fact-finder. Notice here again our central theme of risk inherent in decision as positively associated with the measurement of the evidence and the consequent burden of proof.

In a judicial proceeding in which there is a dispute about the facts of some earlier event, the fact-finder cannot acquire unassailably accurate knowledge of what happened. All the fact-finder can acquire is a belief of what probably happened. The degree to which a fact-finder is convinced that a given act actually occurred can, of course, vary. A standard of proof thus represents an attempt to instruct the fact-finder concerning the degree of confidence society thinks he should have in the correctness of the factual conclusion for a particular type of adjudication. There are phrases such as "preponderance of the evidence" and "proof beyond a reasonable doubt" which are somewhat imprecise yet do communicate to the finder of fact different notions concerning the degree of confidence he is expected to have in the correctness of his factual conclusions. For a class of cases, therefore, the burden of proof lies along a continuum from low probability to very high probability. In civil suits, the "preponderance of the evidence," that is, "more probable than not" standard, is relied upon, that is, the law is indifferent as between plaintiffs and defendants but hopes to minimize the probability of error. On the other hand, equalization of errors between parties may require higher probabilities than minimization of errors. For the first situation, the judge states a quantification of $50 + \%$ probable, and for the latter situation, more than $50 + \%$ probable.

In some civil proceedings, the standard of "clear and convincing evidence" is employed, and this is a test somewhat stricter than preponderance of the evidence. Where proof of another crime is being used as relevant evidence under the Federal rules of evidence, a most common test is some form of the standard, "clear and convincing evidence". In the Fatico case analysis, the judge states that quantified, the probabilities here might be on the order of about 70%, that is, for a "clear and convincing evidence" burden.

In some situations such as deportation proceedings, the Federal Courts require "clear, unequivocal and convincing evidence" for drastic deprivations may follow when a resident of the United States is compelled by the Government to give up all the bonds formed in the United States and go to a foreign land where he often has no contemporary identification. In the Fatico case discussion in which this appears, the judge considers probabilities on the order of about 80% for "clear, unequivocal and convincing evidence" standard.

We now come to the issue of "proof beyond a reasonable doubt" which is constitutionally mandated for elements of a criminal offense. If quantified as it was in the Fatico case, the beyond reasonable doubt standard is considered to be in the range of $95 + \%$ probable.

Judges do not always make fine distinctions between preponderance, clear and convincing, clear, unequivocal and convincing and beyond a reasonable doubt. A number were interviewed (see Table 1 below) and placed the probability standard higher than would be expected on theoretical grounds for a preponderance and somewhat lower than might be expected for beyond a reasonable doubt, thus indicating a narrower range in which to insert the two intermediate burdens. Almost one-third of the responding judges put "beyond a reasonable doubt" at 100%, another third put it at 90 or 95%, and most of the others put it at 80 or 85%. For the preponderance standard, by contrast, over half put it between 60 and 75%. Questionnaires sent to jurors and students produced slightly lower results for the reasonable doubt instruction, and rather high results for the preponderance standard. Yet for most people, the distinction was quite clear. Table 1 emerges from a survey of Federal District Judges in the Eastern District of New York and indicates their assessment of probabilities. This table and a detailed discussion of this subject appears in the article mentioned at the beginning of this section.

Table 1. Probabilities Associated with Standards of Proof (Judges, Eastern District of New York)

Judge	Preponderance (%)	Clear and convincing (%)	Clear, unequivocal and convincing (%)	Beyond a reasonable doubt (%)
1	50+	60–70	65–75	80
2	50+	67	70	76
3	50+	60	70	85
4	51	65	67	90
5	50+	Standard is elusive and unhelpful		90
6	50+	70+	70+	85
7	50+	70+	80+	95
8	50.1	75	75	85
9	50+	60	90	85
10	51	Cannot estimate numerically		

References

Katz, L. (1975). Presentation of a confidence interval estimate as evidence in a legal proceeding, *American Statistician*, **29**, 138–142.

Niblett, B. and Boreham, J. (1976). Cluster analysis in court, *The Criminal Law Review*, **23**, 175–180.

Rubenstein, A. (1979). Fake probabilistic arguments v. faulty instruction, *Israel Law Review*, **14**, 247–254.

Solomon, H. (1966). Jurimetrics. In *Research Paper in Statistics* (F. N. David, ed.), 319–350. Wiley, New York.

Sprowls, R. C. (1957). The admissibility of sample data into a court of law, *UCLA Law Review*, **4**, 233–250.

2

DATA MODELING USING QUANTILE AND DENSITY-QUANTILE FUNCTIONS

Emanuel Parzen

Abstract

Statistical data modeling is a field of statistical reasoning that seeks to fit models to data without using models based on prior theory; rather one seeks to learn the model by a process which could be called statistical model identification. When analyzing a sample $X_1, \ldots, X_n$, statisticians should not confine themselves to either fitting a Gaussian distribution, or transforming the data to be Gaussian. Such an approach ignores the importance of bimodality as a feature of observed data, and also ignores the need to fit to data probability model based distributions which could suggest probability models for the causes generating the data. This paper describes an approach to statistical data modeling which emphasizes estimation of quantile and density-quantile functions; it treats the Gaussian distribution as just one of the available distributions.

Sections 1–3 introduce the role of quantile functions in statistical modeling, the sample quantile function and location and scale parameter models. Quantile function based descriptors of a probability distribution are defined (Section 4). Section 5 defines quantile box plots and transformation distribution functions; an example of their application is discussed in Section 6. A quantile version of "bootstrap" simulation methods is outlined in Section 7. Data summary by a few values of the sample quantile function is discussed (Section 9). Section 8 discusses quantile function formulations of robust estimators of location and scale.

The concepts discussed in this paper are best summarized by a list of some of the terminology defined: quantile function, density-quantile function, score function, sample quantile function, sample quantile-density function, histo-

This research was supported in part by the Army Research Office (Grant DAAG29-80-C-0070).

gram-quantile function, sample entropy, score deviation, tail exponents, mode percentile, quantile box plot, cumulative weighted spacings plot, quantile bootstrap, minimum residual score deviation estimation and 19 quantile values for universal data summary.

1. Some basic concepts of statistical modeling and estimation

One of the basic problems of statistical data analysis is the *one-sample* problem: given a sample $X_1, \ldots, X_n$ which we assume initially to be independent observations of a population characteristic represented by a random variable X, we would like to infer the probability distribution of X.

The probability distribution of X is usually represented by its distribution function

$$F(x) = \Pr[X \leqslant x]$$

and by its probability density function

$$f(x) = F'(x).$$

In this paper we assume X is continuous and possesses a probability density function.

The problem of statistical inference is often defined to be *parameter estimation*; then one assumes that the true probability density function $f(x)$ belongs to a family of functions $f_\theta(x)$ indexed by a vector θ of parameters $\theta_1, \ldots, \theta_r$. The maximum likelihood estimator of θ is defined to be a function $\hat\theta$ of $X_1, \ldots, X_n$ satisfying $L(\hat\theta) = \max_\theta L(\theta)$, defining

$$L(\theta) = f_\theta(X_1, \ldots, X_n) = \prod_{j=1}^{n} f_\theta(X_j);$$

$L(\theta)$ is the joint probability density of the observed data when θ is the true parameter value.

Maximum likelihood estimation is not a principle to be accepted uncritically; statisticians delight in constructing examples which lead to unbelievable conclusions. To understand when and why maximum likelihood estimation works, we have to introduce the empirical distribution function (EDF) $\tilde F(x)$ defined by

$$\tilde F(x) = \text{fraction of } X_1, \ldots, X_n \leqslant x.$$

To graph $\tilde F(x)$, one determines the order statistics $X_{(1)} \leqslant X_{(2)} < \cdots < X_{(n)}$, which are the sample values (assumed to be distinct) arranged in increasing

2 DATA MODELING USING QUANTILE AND DENSITY-QUANTILE FUNCTIONS

order; then

$$\tilde{F}(x) = \frac{j}{n}, \qquad X_{(j)} \leqslant x < X_{(j+1)}, \qquad j = 0, 1, 2, \ldots, n,$$

where $X_{(0)} = -\infty$ and $X_{(n+1)} = \infty$.

The concept of likelihood is now defined as average log likelihood

$$L_n(\theta) = \frac{1}{n} \log \prod_{j=1}^{n} f_\theta(X_j)$$

$$= \frac{1}{n} \sum_{j=1}^{n} \log f_\theta(X_j)$$

$$= \int_{-\infty}^{\infty} \log f_\theta(x) \, d\tilde{F}(x).$$

One can regard $L_n(\theta)$ as a measure of "distance" between the data represented by $\tilde{F}$, and the model represented by $f_\theta(x)$.

Another important interpretation of $L_n(\theta)$ is an *estimator* of a "distance" between the true probability density $f(x)$ and the model $f_\theta(x)$. An important role in the theory of statistical inference is played by the Kullback–Liebler information number, or directed divergence (Zacks, 1971); it is defined by

$$I(f; f_\theta) = E_f \left[\log \frac{f}{f_\theta} \right]$$

$$= \int_{-\infty}^{\infty} f(x) \log \frac{f(x)}{f_\theta(x)} \, dx$$

$$= H(f; f) - H(f; f_\theta)$$

defining

$$H(f; g) = \int_{-\infty}^{\infty} f(x) \log g(x) \, dx.$$

It has the properties: $I(f; f_\theta) \geqslant 0$ and $I(f; f) = 0$. The average directed divergence between f and f_θ given a sample $X_1, \ldots, X_n$ is

$$I_n(f; f_\theta) = \frac{1}{n} E_f \log \frac{f(X_1, \ldots, X_n)}{f_\theta(X_1, \ldots, X_n)}$$

$$= \frac{1}{n} \int_{-\infty}^{\infty} \cdots \int_{-\infty}^{\infty} f(x_1, \ldots, x_n) \log \frac{f(x_1, \ldots, x_n)}{f_\theta(x_1, \ldots, x_n)} \, dx_1, \ldots, dx_n$$

$$= I(f; f_\theta).$$

A criterion for model fitting is to choose f_θ to minimize $I(f;f_\theta)$ or an estimator of $I(f;f_\theta)$; an estimator would be

$$I(\tilde{f};f_\theta) = H(\tilde{f};\tilde{f}) \to H(\tilde{f};f_\theta)$$

if $\tilde{f}$ were a non-parametric estimator of f. While $\tilde{F}$ is a natural non-parametric estimator of F, there does not exist a natural non-parametric estimator of f. However a natural non-parametric estimator $\tilde{H}(f;f_\theta)$ of $H(f;f_\theta)$ does exist, namely the average log likelihood $L_n(\theta)$; in symbols,

$$\tilde{H}(f;f_\theta) = \int_{-\infty}^{\infty} \log f_\theta(x)\,d\tilde{F}(x).$$

A natural estimator $\tilde{H}(f;f)$ will be given below. Akaike (1973) has pioneered in emphasizing that to find $\hat{\theta}$, the parameter values θ which minimize

$$\tilde{I}(f;f_\theta) = \tilde{H}(f;f) - \tilde{H}(f;f_\theta),$$

it is not necessary to know $\tilde{H}(f;f)$; one need only choose $\hat{\theta}$ to maximize $L_n(\theta)$. One approach to measuring how well the maximum likelihood model $f_{\hat\theta}$ "matches" the data would be to measure how significantly different from zero is $\tilde{I}(f;f_{\hat\theta})$. Other approaches to measuring the mathematical fit of a model to data are introduced in this paper using various representing functions of the data and model which are called the "raw" and "smooth" representing functions, respectively. One of our goals is to develop means of judging goodness of fit of a family f_θ of probability densities to a true probability density *before* forming estimators $\hat\theta$ of the parameters.

This paper discusses the increased insight to be obtained by describing the probability distribution of a random variable X by its *quantile* function $Q(u)$, $0 \leq u \leq 1$, and *density-quantile* function $fQ(u)$, $0 \leq u \leq 1$. Define

$$Q(u) = F^{-1}(u) = \inf\{x: F(x) \geq u\},$$
$$fQ(u) = f[Q(u)].$$

The *quantile-density* function $q(u)$, $0 \leq u \leq 1$, is the derivative of the quantile function:

$$q(u) = Q'(u).$$

The *score* function is (-1) times the derivative of the density-quantile function:

$$J(u) = -(fQ)'(u).$$

An important identity is

$$fQ(u)q(u) = 1,$$

2 DATA MODELING USING QUANTILE AND DENSITY-QUANTILE FUNCTIONS

which follows by differentiating the identity

$$FQ(u) = u.$$

We can now give an example of the advantages of thinking "quantile" in the sense of thinking in terms of $fQ(u)$ rather than $f(x)$. Two measures of the smoothness of a function are the integral of its logarithm and the integral of its derivative squared. Thus

$$\int_0^1 \log fQ(u) = \int_{-\infty}^{\infty} f(x)\log f(x)\,dx = H(f;f)$$

is the Shannon information measure or entropy of f, while the Fisher information measure of f is

$$\int_0^1 |J(u)|^2\,du = \int_0^1 |(fQ)'(u)|^2\,du = \int_0^1 \left|\frac{f'Q(u)}{fQ(u)}\right|^2 du = \int_{-\infty}^{\infty} \frac{|f'(x)|^2}{f(x)}\,dx.$$

One can give a natural estimator of entropy:

$$\tilde{H}(f;f) = -\int_0^1 \log \tilde{q}(u)\,du,$$

where $\tilde{q}(u)$ is the sample quantile-density defined below. We call $\tilde{H}$ the *sample entropy*.

The density-quantile function as a function of interest for itself was introduced by Parzen (1979a). Tukey (1965) pointed out the significance of $Q(u)$ and $q(u)$ under the names "representing" function and "sparsity" function. A review of some standard approaches to statistical modeling is given by Ord and Patil (1975).

2. Sample quantile function

To a batch of data one can define a sample quantile function $\tilde{Q}(u)$, $0 \leq u \leq 1$, which provides a "universal" description and summary of the data. However, there is no universally accepted definition of $\tilde{Q}(u)$.

Given a sample $X_1, \ldots, X_n$, with order statistics $X_{(1)} < X_{(2)} < \cdots < X_{(n)}$ one could define $\tilde{Q}$ by

$$\tilde{Q}(u) = \tilde{F}^{-1}(u), \qquad 0 \leq u \leq 1;$$

then $\tilde{Q}$ is piecewise constant,

$$\tilde{Q}(u) = X_{(j)}, \qquad \frac{j-1}{n} < u \leq \frac{j}{n}, \quad j = 1, 2, \ldots, n.$$

One often prefers a piecewise-linear definition of $\tilde{Q}(u)$; then one defines

$$\tilde{Q}(u) = X_{(j)} \quad \text{if } u = \frac{j-0.5}{n} \text{ or } \frac{j}{n+1}.$$

One also defines values for $u = 0$ or 1, say $\tilde{Q}(0) = X_{(1)}$ and $\tilde{Q}(1) = X_{(n)}$. At other values of u in $0 < u < 1$, one defines $\tilde{Q}(u)$ by linear interpolation of its values at the grid points $(j-0.5)/n$ or $j/(n+1)$. Then $\tilde{Q}(u)$ is differentiable, and $\tilde{q}(u) = \tilde{Q}'(u)$ may be expressed in terms of the sample "spacings"

$$n\{X_{(j+1)} - X_{(j)}\}.$$

When $\tilde{Q}(u) = X_{(j)}$ at $u = (j-0.5)/n$, then $(j = 1, 2, \ldots, n-1)$,

$$\tilde{q}(u) = n\{X_{(j+1)} - X_{(j)}\}, \quad \frac{j-0.5}{n} < u < \frac{j+0.5}{n}.$$

A favorite tool of statistical data analysis is the histogram which can be defined as a piecewise-constant estimator $\tilde{f}(x)$ of the density function. The sample quantile function $\tilde{Q}(u)$ is then defined as the inverse of the sample distribution function $\tilde{F}(x) = \int_{-\infty}^{x} \tilde{f}(y)\,dy$. The insight in a histogram seems to me to be made more visible by plotting instead the *histogram-quantile* function $\tilde{f}[\tilde{Q}(u)], 0 \leq u \leq 1$.

A raw estimator of $fQ(u)$, called a raw density-quantile function and denoted $\tilde{f}Q(u)$, can be formed from the reciprocal of a slightly smoothed estimator of $q(u)$; for example, one might define

$$\tilde{f}Q(u) = \frac{2h}{\tilde{Q}(u+2h) - \tilde{Q}(u-2h)}.$$

The sample quantile function $\tilde{Q}(u), 0 \leq u \leq 1$, is a stochastic process (or time series) whose asymptotic distribution can be shown to satisfy (under suitable assumptions on fQ (Csorgo and Revesz, 1978)).

$$\{\sqrt{n}fQ(u)\{\tilde{Q}(u) - Q(u)\}, 0 \leq u \leq 1\} \to \stackrel{L}{=} \{B(u), 0 \leq u \leq 1\},$$

where $\{B(u), 0 \leq u \leq 1\}$, denotes a Brownian Bridge stochastic process with covariance function

$$E[B(u_1)B(u_2)] = u_1(1-u_2) \quad \text{for } 0 \leq u_1 \leq u_2 \leq 1,$$

$\stackrel{L}{=}$ denotes "identically distributed as", and the convergence is in the sense of convergence of distribution of stochastic processes.

The asymptotic distribution of the sample spacings, and thus of $\tilde{q}(u)$, also have been extensively investigated but is difficult to summarize briefly. One important fact is that for any fixed $u_1, \ldots, u_k$

$$fQ(u_1)\tilde{q}(u_1), \ldots, fQ(u_k)\tilde{q}(u_k)$$

are asymptotically independent and distributed as an exponential distribution with mean 1.

The difference between the roles of distribution functions and quantile functions in statistical inference is made clear by considering the basic goodness-of-fit problem: test the hypothesis H_0,

$$H_0: F(x) = F_0(x), \quad -\infty < \infty < \infty,$$

that the true distribution function $F(x)$ equals a specified distribution function $F_0(x)$. One could compare the sample distribution $\tilde{F}(x)$ to $F_0(x)$ (or equivalently test whether the transformed random variables $F_0(X_1), \ldots, F_0(X_n)$ are uniformly distributed) by comparing $\tilde{F}[Q_0(u)]$ to u. The applicable asymptotic distribution theorem is

$$\{\sqrt{n}\{\tilde{F}[Q_0(u)] - u\}, 0 \leqslant u \leqslant 1\} \xrightarrow{L} \{B(u), 0 \leqslant u \leqslant 1\}.$$

Alternately one could compare quantile functions. Instead of comparing $\tilde{Q}(u)$ to $Q_0(u) = F_0^{-1}(u)$, one could *compare the sample quantile function of* $F_0(X_1), \ldots, F_0(X_n)$, which equals $F_0[\tilde{Q}(u)]$, to u. The relevant asymptotic distribution theorem is

$$\{\sqrt{n}\{F_0[\tilde{Q}(u)] - u\}, 0 \leqslant u \leqslant 1\} \xrightarrow{L} \{B(u), 0 \leqslant u \leqslant 1\}.$$

The problem of statistical modeling can be elegantly defined in terms of quantile functions: one seeks to determine distribution functions $F_0(x)$ such that $F_0[\tilde{Q}(u)]$ is not significantly different from a uniform quantile function u. Given a parametric family of distribution functions $F_\theta(x)$ an optimal estimator $\hat{\theta}$ of θ could be defined as the value of θ which minimizes the distance $\|F_\theta[\tilde{Q}(u)] - u\|$ for a suitable measure of distance between functions on the interval 0 to 1.

An example of a distance is the conventional L_2 distance

$$\|g_1 - g_2\|^2 = \int_0^1 |g_1(u) - g_2(u)|^2 \, du.$$

However, one would like to choose the distance so that the estimator $\hat{\theta}$ would be asymptotically efficient. Such a distance is provided by the *reproducing kernel Hilbert space* (RKHS) norm of the covariance kernel of the Brownian Bridge stochastic process; it can be defined over any subinterval $0 \leqslant p < u \leqslant q < 1$:

$$\|g_1 - g_2\|^2_{p,q} = \int_p^q |g_1'(u) - g_2'(u)|^2 \, du + \frac{1}{p}|g_1(p) - g_2(p)|^2$$

$$+ \frac{1}{1-q}|g_1(q) - g_2(q)|^2,$$

$$\|g_1 - g_2\|_{0,1}^2 = \int_0^1 |g_1'(u) - g_2'(u)|^2 \, du.$$

The inner product is

$$(g_1, g_2)_{p,q} = \int_p^q g_1'(u) g_2'(u) \, du + \frac{1}{p} g_1(p) g_2(p) + \frac{1}{1-q} g_1(q) g_2(q).$$

A minimum distance criterion for statistical estimation of the parameters θ of a parametric family F_θ of distribution functions is to choose θ to minimize

$$\|F_\theta[\tilde{Q}(u)] - u\|^2 = \int_0^1 |f_\theta[\tilde{Q}(u)]\tilde{q}(u) - 1|^2 \, du.$$

One may show this criterion to be asymptotically equivalent to maximizing likelihood, or minimizing the estimated directed divergence $I(\tilde{f}, f_\theta)$:

$$I(\tilde{f}; f_\theta) = \int_{-\infty}^{\infty} \tilde{f}(x) \log \frac{\tilde{f}(x)}{f_\theta(x)} \, dx$$

$$= -\int_0^1 \log\{f_\theta[\tilde{Q}(u)]\tilde{q}(u)\} \, du.$$

3. Location and scale parameter models

An important parametric model for a distribution function $F(x)$ is

$$F(x) = F_0\left(\frac{x - \mu}{\sigma}\right),$$

where F_0 is specified, and μ and σ are unknown (location and scale) parameters to be estimated. Then

$$Q(u) = \mu + \sigma Q_0(u),$$
$$q(u) = \sigma q_0(u),$$
$$fQ(u) = \frac{1}{\sigma} f_0 Q_0(u).$$

Two important choices for F_0 are:

(1) The normal or Gaussian case:

$$F_0(x) = \Phi(x) = \int_{-\infty}^x \phi(y) \, dy,$$

$$f_0(x) = \phi(x) = \frac{1}{\sqrt{2\pi}} \exp(-\tfrac{1}{2} x^2).$$

2 DATA MODELING USING QUANTILE AND DENSITY-QUANTILE FUNCTIONS

Table 1. Quantile Functions, Score Functions and Density-Quantile Functions

Probability law	$Q(u)$	$J(u)$	$fQ(u)$
Normal	$\Phi^{-1}(u)$	$\Phi^{-1}(u)$	$\phi\Phi^{-1}(u) = (2\Pi)^{-1/2}\exp\{-\frac{1}{2}[\Phi^{-1}(u)]^2\}$
Log-normal	$\exp[\Phi^{-1}(u)]$	$\exp[-\Phi^{-1}(u)]\{\Phi^{-1}(u)+1\}$	$\phi\Phi^{-1}(u)\exp[-\Phi^{-1}(u)]$
Exponential	$\log(1-u)^{-1}$	1	$1-u$
Extreme value	$\log\log(1-u)^{-1}$	$1+\log(1-u)^{-1}$	$(1-u)\log(1-u)^{-1}$
Weibull	$\{\log(1-u)^{-1}\}^{\beta}$	$\{\log(1-u)^{-1}\}^{-\beta}\{1-\beta+\log(1-u)^{-1}\}$	$(1/\beta)(1-u)\{\log(1-u)^{-1}\}^{1-\beta}$
Logistic	$\log u/(1-u)$	$2u-1$	$u(1-u)$
Double-exponential	$\log 2u, u < \frac{1}{2}$ $-\log 2(1-u), u > \frac{1}{2}$	$\text{sign}(2u-1)$	$u, u < 0.5; 1-u, u > 0.5$
Cauchy	$\tan\Pi(u-\frac{1}{2})$	$-\sin 2\Pi u$	$(1/\Pi)\sin^2\Pi u$
Pareto	$(1/\beta)(1-u)^{-\beta}$	$(1+\beta)(1-u)^{\beta}$	$(1/\beta)(1-u)^{1+\beta}$

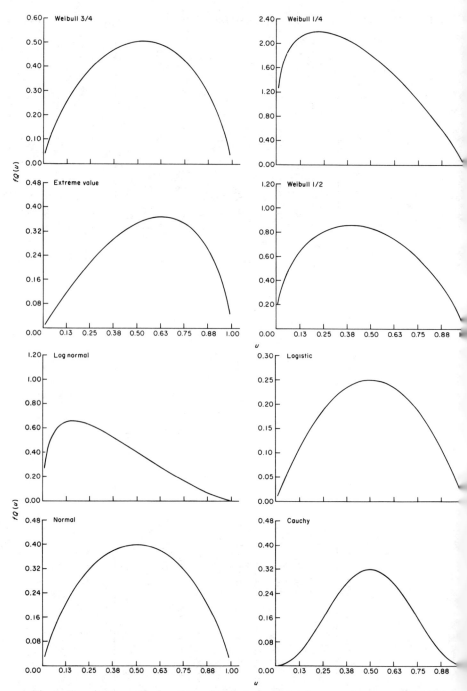

Fig. 1. Density quantile functions $fQ(u)$, $0 \leq u \leq 1$, of some common probability distributions: log-normal, logistic, normal, Cauchy, Weibull with various shape parameters and extreme value.

2 DATA MODELING USING QUANTILE AND DENSITY-QUANTILE FUNCTIONS

(2) The exponential case:

$$F_0(x) = 1 - e^{-x}, \qquad f_0(x) = e^{-x}.$$

The quantile functions, score functions and density-quantile functions of some standard probability laws are given in Table 1. Graphs of density-quantile functions are given in Fig. 1.

Because of the way that $fQ(u)$ depends on μ and σ, one can introduce functions to test hypothesis $H_0: Q(u) = \mu + \sigma Q_0(u)$ which do not require estimation of μ and σ before testing the hypothesis. Define

$$\sigma_0 = \int_0^1 f_0 Q_0(u) q(u) \, du,$$

$$d(u) = \frac{1}{\sigma_0} f_0 Q_0(u) q(u),$$

$$D(u) = \int_0^u d(u') \, du'.$$

We call $D(u)$ a *transformation distribution* function, and $d(u)$ a *transformation density*. The null hypothesis H_0 is equivalent to

$$D(u) = u, \qquad d(u) = 1, \qquad 0 \leq u \leq 1.$$

Given an estimator $\tilde{D}(u), 0 \leq u \leq 1$, the deviations of $\tilde{D}(u)$ from linearity can be used to test whether a sample consists of random variables satisfying H_0, or consists of random variables satisfying H_0 plus outliers, Such techniques would be useful for many diverse applications.

4. Quantile based measures of average, deviation, tail behavior and modes

We propose that the sample quantile function provides a *representing function* for the sample in the following senses:

(1) Models for the data should be viewed as being in one to one correspondence with the smooth quantile functions $\hat{Q}(u), 0 \leq u \leq 1$, which are their representing functions.
(2) The criteria for testing whether a model fits the data, should be based on measures of fit between the representing functions $\hat{Q}(u)$ and $\tilde{Q}(u)$.
(3) Since the sample is summarized by its representing function $\tilde{Q}(u)$, any descriptor of the sample should be expressible as a function of $\tilde{Q}(u)$. Similarly any descriptor of the distribution of X should be expressible as a function of $Q(u)$.

There are four characteristics of a probability distribution which we would like to infer from the data:

(1) *Location*, represented by a measure of average.
(2) *Spread*, represented by a measure of deviation.
(3) *Tail behavior*, represented by the behavior of $fQ(u)$ as u tends to 0 and 1.
(4) *Modality*, represented by the number of modes (relative maximum) in the probability density or in the density-quantile function.

Location and spread parameters for a distribution seem to be meaningful only when it is unimodal; otherwise, one may want to find an associated variable whose values can be used to divide the original sample of X values into two or more samples, each of which is unimodal.

A parameter representing average or location will be denoted by μ; it could be defined by one of the following concepts:

$$\text{median } \mu = Q(0.5),$$

$$\text{mid-quartile } \mu = \tfrac{1}{2}\{Q(0.25)+Q(0.75)\},$$

$$\text{mid-range } \mu = \tfrac{1}{2}\{Q(0)+Q(1)\},$$

$$\text{mean } \mu = \int_0^1 Q(u)\,du = \int_{-\infty}^{\infty} xf(x)\,dx.$$

A parameter representing deviation or spread or scale will be denoted by σ; it could be defined by one of the following concepts:

$$\text{interquartile range } \sigma = Q(0.75)-Q(0.25),$$

$$\text{score deviation } \sigma = \int_0^1 J_0(u)Q(u)\,du \text{ with score function } J_0(u),$$

$$\text{standard deviation } \sigma = \left\{\int_0^1 \left\{Q(u)-\int_0^1 Q(t)\,dt\right\}^2 du\right\}^{1/2}.$$

The properties of $fQ(u)$ describe the tail behavior, modality and symmetry of the distribution. Indices α_1 and α_2 such that

$$fQ(u) \sim u^{\alpha_1} \quad \text{as } u \to 0,$$

$$fQ(u) \sim (1-u)^{\alpha_2} \quad \text{as } u \to 1,$$

may be rigorously defined when they exist by

$$\alpha_1 = \lim_{u \to 0} \frac{-uJ(u)}{fQ(u)}, \quad \alpha_2 = \lim_{u \to 1} \frac{(1-u)J(u)}{fQ(u)}.$$

2 DATA MODELING USING QUANTILE AND DENSITY-QUANTILE FUNCTIONS

We call α_1 the left tail exponent, and α_2 the right tail exponent.

The tail exponent α indicates whether the tail is short, medium, or long: $\alpha < 1$, *short;* $\alpha = 1$, *medium;* $\alpha > 1$, *long.*

The Gaussian distribution has $\alpha_1 = \alpha_2 = \alpha = 1$; the exponential distribution has $\alpha_1 = 0$ and $\alpha_2 = 1$; the Cauchy distribution has $\alpha_1 = \alpha_2 = \alpha = 2$. The graphs of their fQ functions are given in Fig. 1. Our ideas of the canonical shapes of distributions seem to me to become unified when they are formulated not in terms of the shape of $f(x)$ but in terms of the shape of $fQ(u)$; for example, J and U-shaped distributions correspond to $\alpha \leq 0$.

When $fQ(u)$ is unimodal, an important descriptor is the *mode percentile*, denoted p_{mode}. It is defined to be the value of u at which $fQ(u)$ achieves its mode (or maximum value). The value of p_{mode} and its relation to 0.5, is a quick summary of the skewness of the distribution.

When p-mode ≥ 0.5, the distribution is conventionally described as being skewed to the left; this occurs if we assume that fQ satisfies $fQ(u) \leq fQ(1-u)$, $0 \leq u \leq 0.5$, which implies that $Q(u) + Q(1-u) \leq 2Q(0.5)$, and consequently that

$$\text{mean} \leq \text{median} \leq \text{mode}.$$

Similarly p-mode ≤ 0.5 (and the distribution is skewed to the right) if we assume that $fQ(u) \geq fQ(1-u)$, $0 \leq u < 0.5$, which implies that $Q(u) + Q(1-u) \geq 2Q(0.5)$, and consequently that

$$\text{mean} \geq \text{median} \geq \text{mode}.$$

The fact that a density-quantile function is always defined on the unit interval, while a density function $f(x)$ is defined on an infinite interval, seems to make the former easier to estimate.

5. Quantile box plots and transform distribution functions

The most dramatic new data-analytic tools suggested by the quantile and density-quantile approach are quantile box plots of $\tilde{Q}(u), 0 \leq u \leq 1$, and plots of sample transformation distribution functions $\tilde{D}(u), 0 \leq u \leq 1$.

Quantile box plots are formed of the original data and of the data after transformations such as square root, logarithm and reciprocal. They provide quick procedures for estimating location, scale and shape. A quantile box plot consists of a graph of a quantile function on which is superimposed various boxes with vertices $[p, \tilde{Q}(p)]$, $[p, \tilde{Q}(1-p)]$, $[1-p, \tilde{Q}(p)]$, $[1-p, \tilde{Q}(1-p)]$, which we call a p-box. One usually chooses $p = \frac{1}{4}, \frac{1}{8}, \frac{1}{16}$. Within the quantile box ($p = 0.25$), one draws a median line with vertices $[0.25, \tilde{Q}(0.5)]$, $[0.75 \tilde{Q}(0.5)]$. An approximate confidence interval for the median $\tilde{Q}(0.5)$ is indicated by a vertical line with vertices $[0.5, \tilde{Q}(0.5) \pm IQ/\sqrt{n}]$ where n is the sample size and $IQ = \tilde{Q}(0.75) - \tilde{Q}(0.25)$ is the inter-quartile range. The symmetry of the distribution is judged by the symmetry of $\tilde{Q}(u)$ within the quartile box.

A quantile box plot is an extension of the idea of a box plot introduced by Tukey (1977).

A transformation distribution function, or cumulative weighted spacings function, is defined by

$$\tilde{D}(u) = \int_0^u \tilde{d}(t)\,dt, \quad 0 \leq u \leq 1,$$

where

$$\tilde{d}(u) = \frac{1}{\tilde{\sigma}_0} f_0 Q_0(u) \tilde{q}(u), \quad \tilde{\sigma}_0 = \int_0^1 f_0 Q_0(u) \tilde{q}(u)\,du.$$

Its pseudo-correlations are defined by

$$\tilde{\rho}(v) = \int_0^1 e^{2\pi i u v} \tilde{d}(u)\,du, \quad v = 0, \pm 1, \ldots.$$

The asymptotic distributional properties of $\tilde{d}(u)$ are similar to those of the sample spectral density of a stationary time series. Tests of H_0 could be based on $\int_0^1 \log \tilde{d}(u)\,du$; the deviation from $D(u) = u$ of $\tilde{D}(u)$; the deviation from $\rho(v) = 0$ of $\tilde{\rho}(v), v = 1, 2, \ldots$.

To estimate the density-quantile function $fQ(u)$, one uses $\tilde{d}(u)$ to form smooth estimators $\hat{d}(u)$ of $d(u)$. Two main approaches are:

(1) Kernel method—for a suitable kernel K

$$\hat{d}_K(u) = \int_0^1 \tilde{d}(t) K(u-t)\,dt.$$

(2) Autoregressive method—for a suitable order m

$$\hat{d}_m(u) = \hat{\sigma}_m^2 |1 + \hat{\alpha}_m(1) e^{2\pi i u} + \cdots + \hat{\alpha}_m(m) e^{2\pi i u m}|^{-2},$$

where $\hat{\sigma}_m^2, \hat{\alpha}_m(j), j = 1, \ldots, m$, are determined from certain linear equations (Yule–Walker equations) in

$$\hat{\rho}(v) = \int_0^1 e^{2\pi i u v} \tilde{d}(t)\,du, \quad v = 0, \pm 1, \ldots, \pm m.$$

The autoregressive estimator, including procedures for selecting the order m, are implemented in a computer program ONESAM whose use is illustrated. It should be noted that choosing order $m = 0$ is equivalent to accepting H_0.

A solution to the important problem of estimating $fQ(u)$ is provided by the "autoregressive" estimator

$$\hat{f}Q_m(u) = f_0 Q_0(u)\{\tilde{\sigma}_0 \hat{d}_m(u)\}^{-1}.$$

Some diagnostics we use for choosing the order m of the autoregressive estimator $\hat{f}Q_m(u)$ are the square modulus pseudo-correlations $|\hat{\rho}(v)|^2$; the residual variances $\hat{\sigma}_m$; the Akaike order determining criterion, for sample size T,

$$\text{AIC}(m) = \log \hat{\sigma}_m^2 + \frac{2m}{T};$$

and Parzen's criterion

$$\text{CAT}(m) = \frac{1}{T} \sum_{j=1}^{m} \hat{\sigma}_j^{-2} - \hat{\sigma}_m^{-2}$$

whose shape in practice is similar to the shape of AIC.

Another approach to estimating $fQ(u)$ which deserves investigation is to estimate $\log fQ(u)$ by smoothing $-\log \tilde{q}(u)$.

6. Examples

To illustrate how to use Q, D and fQ in statistical data analysis, let us consider data from Tukey (1977), p. 117, which lists seasonal snowfall in Buffalo, New York, and Cairo, Illinois, from 1918–19 to 1937–38, and asks "What light do these two batches throw on how they should be expressed?" To answer this question one approach might be to examine the quantile box plots of the batches (Fig. 2); the quartile box in Buffalo appears symmetric while in Cairo it does not. One might attempt a transformation of the Cairo data; we choose the square root and conclude that its quartile box is symmetric. Does this prove that Buffalo snowfall and square root of Cairo snowfall are Gaussian?

A rigorous approach is to form the cumulative weighted spacings function

$$\tilde{D}(u) = \frac{\int_0^u \phi \Phi^{-1}(t) \tilde{q}(t) \, dt}{\int_0^1 \phi \Phi^{-1}(t) \tilde{q}(t) \, dt}, \quad 0 \leq u \leq 1,$$

whose deviations from u provide a test of Gaussian-ness which does not first require estimation of μ and σ (mean and standard deviation). The graphs of $\tilde{D}(u)$ in Fig. 3 indicate clearly that Buffalo snowfall and square root of Cairo are Gaussian, while Cairo snowfall is not Gaussian.

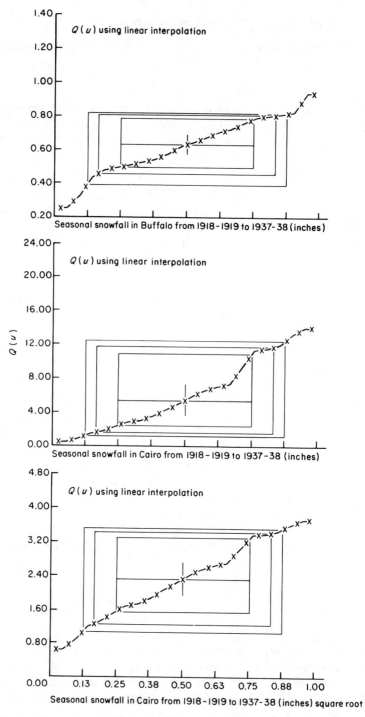

Fig. 2. Quantile box plots of Buffalo and Cairo and square root Cairo snowfall.

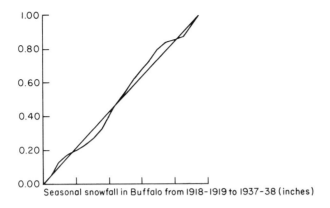
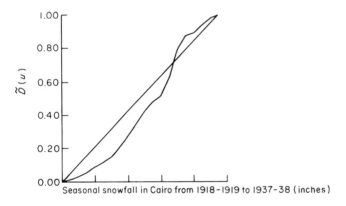
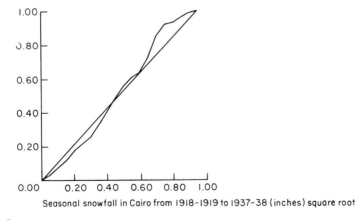

Fig. 3. $\tilde{D}(u)$ curves for testing normality.

Table 2. Buffalo and Cairo Snowfall

Order statistics	Buffalo	Cairo	Cairo square root
1	25.0	0.4	0.6325
2	39.8	1.2	1.0954
3	46.7	1.6	1.2649
4	49.1	1.8	1.3416
5	49.6	2.7	1.6432
6	51.6	2.9	1.7029
7	53.5	3.0	1.7320
8	54.7	4.0	2.0000
9	60.3	4.5	2.1213
10	63.6	5.4	2.3238
11	64.8	6.2	2.4900
12	69.4	6.8	2.6077
13	71.8	7.2	2.6833
14	72.9	7.4	2.7203
15	79.0	11.3	3.3615
16	79.6	11.5	3.3912
17	80.7	11.5	3.3912
18	81.6	12.4	3.5214
19	83.6	13.9	3.7283
20	103.9	14.1	3.7550
Mean $\hat{\mu}$	64.1	6.5	2.375
$\tilde{Q}(0.25)$	49.6	2.7	1.6432
$\tilde{Q}(0.50)$	64.2	5.8	2.4069
$\tilde{Q}(0.75)$	79.6	11.5	3.3912
IQ	30.0	8.8	1.748
SD $\hat{\sigma}$	18.4	4.5	0.945

The true character of the Cairo snowfall data emerges when one estimates its fQ function; it turns out to be bimodal, which we interpret to mean that there are two kinds of snowfall years in Cairo, Illinois—light and heavy.

Even though various diagnostic tests of the square roots of Cairo snowfall data (Table 2) indicate that it is a Gaussian, the order 1 autoregressive estimator of the density quantile (Table 3 and Fig. 4) indicates that bimodality is a possible alternative hypothesis.

7. Quantile simulation and quantile bootstrap

The quantile function $Q(u)$, $0 \leq u \leq 1$, of a random variable X provides a way of simulating a random sample of X. Let U denote a random variable uniform

Table 3. Buffalo and Cairo Snowfall Autoregressive Analysis Diagnostics

	Buffalo	Cairo	Cairo square root
	$\|\hat{\rho}(v)\|^2$		
0	1.0000	1.0000	1.0000
1	0.0203	0.1391	0.0705
2	0.0187	0.0192	0.0142
3	0.0053	0.0215	0.0222
4	0.0196	0.0496	0.0289
5	0.0169	0.0317	0.0181
	$\sigma^2(m)$		
0	1.0000	1.0000	1.0000
1	0.9797	0.8609	0.9295
2	0.9657	0.8199	0.8963
3	0.9559	0.8179	0.8603
4	0.9367	0.7883	0.8570
5	0.9021	0.7610	0.8404
	AIC(m)		
0	0.0000	0.0000	0.0000
1	0.0795	−0.0497	0.0269
2	0.1651	0.0014	0.0905
3	0.2549	0.0990	0.1495
4	0.3346	0.1622	0.2457
5	0.3969	0.2269	0.3261
Minimum at $m =$	0	1	0
	CAT		
0	−1.0000	−1.0000	−1.0000
1	−0.9212	−1.0483	−0.9709
2	−0.8369	−0.9877	−0.9028
3	−0.7497	−0.8772	−0.8373
4	−0.6718	−0.8020	−0.7361
5	−0.6076	−0.7235	−0.6504

on 0 to 1; then $X \stackrel{L}{=} Q(U)$. Let $U_1,\ldots,U_n$ be independent uniform random variables; then

$$X_1,\ldots,X_n \stackrel{L}{=} Q(U_1),\ldots,Q(U_n).$$

To generate a random sample $X_1,\ldots,X_n$, one generates n random numbers $U_1,\ldots,U_n$ and transforms them to $X_1,\ldots,X_n$.

To obtain by Monte Carlo methods the distribution of a statistic

$$T = g(X_1,\ldots,X_n),$$

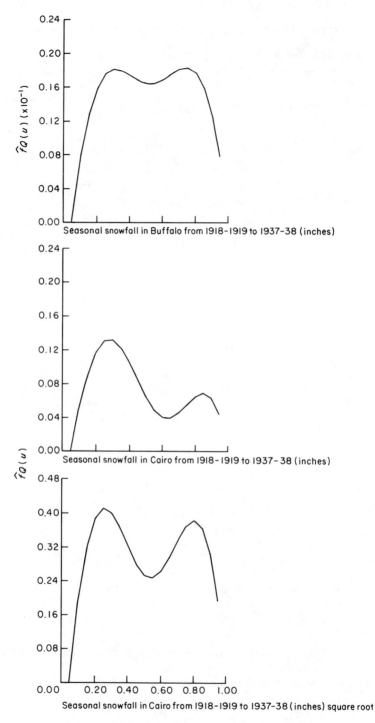

Fig. 4. Autoregressive order 1 estimates of density-quantile function.

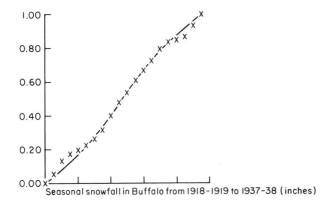
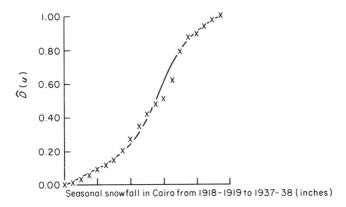
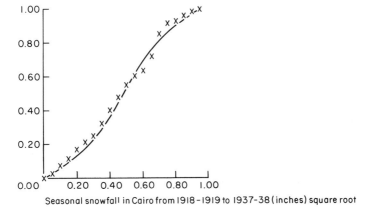

Fig. 4 (*cont.*). $\hat{D}(u)$, solid line, is compared with $\tilde{D}(u)$, crosses.

one would generate a large number of N of random samples $X_1, \ldots, X_n$; generate a random sample $T_1, \ldots, T_N$ of the random variable T; and finally form the sample quantile function $\tilde{Q}_T(u)$ which provides an estimator of the true quantile function of T.

When the quantile function $Q(u)$ of X is not known, one can estimate it by the sample quantile function $\tilde{Q}(u)$. Now from random numbers $U_1, \ldots, U_n$, one can generate "bootstrap" simulated values (cf. Efron, 1978)

$$\tilde{X}_1 = \tilde{Q}(U_1), \ldots, \tilde{X}_n = \tilde{Q}(U_n), \qquad \tilde{T} = g(\tilde{X}_1, \ldots, \tilde{X}_n).$$

One can generate a random sample $\tilde{T}_1, \ldots, \tilde{T}_N$ of T, whose sample quantile function $\tilde{Q}_{\tilde{T}}(u)$ provides an estimator of the true quantile function of T.

Bivariate distributions

An outstanding problem of statistics is the simulation of multivariate distributions. To illustrate the quantile approach to this problem, let (X_1, X_2) have joint distribution function $F(x_1, x_2)$. Denote the marginal distributions functions of X_1 and X_2 by $F_1(x_1)$ and $F_2(x_2)$. Denote the quantile functions of the marginal distributions by $Q_1(u_1)$ and $Q_2(u_2)$. Define

$$D(u_1, u_2) = F[Q_1(u_1), Q_2(u_2)];$$

it is the joint distribution function of the "rank transforms"

$$U_1 = F_1(X_1), \qquad U_2 = F_2(X_2).$$

One can generate (X_1, X_2) by generating (U_1, U_2) from the distribution $D(u_1, u_2)$ and then forming

$$X_1 = Q_1(U_1), \qquad X_2 = Q_2(U_2).$$

To generate (U_1, U_2) one chooses U_1 to be uniform on 0 to 1, and then generates U_2 by the conditional distribution $D_{U_2|U_1}(u_2|u_1)$ or its quantile function $Q_{U_2|U_1}(p|u_1)$ by the formula $(U_2|U_1 = u_1) \stackrel{L}{=} Q_{U_2|U_1}(U_2'|u_1)$, where U_2' is uniform on 0 to 1, and independent of U_1. The conditional quantile function $Q_{U_2|U_1}(p|u_1)$ can be estimated by the sample quantile function of the sample values U_2 corresponding to sample U_1 values near u_1.

8. Quantile formulation of robust location and scale estimators

Assume a location-scale parameter model for the quantile function of a continuous random variable X: $Q(u) = \mu + \sigma Q_0(u)$. Assume a symmetric distribution, which is equivalent to $Q_0(u)$ being an odd function in the sense that $Q_0(1-u) = -Q_0(u)$.

2 DATA MODELING USING QUANTILE AND DENSITY-QUANTILE FUNCTIONS

Given a sample $X_1, \ldots, X_n$, the log likelihood function may be written in terms of the sample quantile function (compare Parzen, 1979b):

$$L = \frac{1}{n} \log f(X_1, \ldots, X_n; \mu, \sigma)$$

$$= \frac{1}{n} \sum_{i=1}^{n} \log \frac{1}{\sigma} f_0\left(\frac{X_i - \mu}{\sigma}\right)$$

$$= -\log \sigma + \int_{-\infty}^{\infty} \log f_0\left(\frac{x - \mu}{\sigma}\right) d\tilde{F}(x)$$

$$= -\log \sigma + \int_0^1 \log f_0\left(\frac{\tilde{Q}(u) - \mu}{\sigma}\right) du. \quad (8.1)$$

The maximum likelihood estimators $\hat{\mu}$ and $\hat{\sigma}$ satisfy $\partial L/\partial \mu = 0$ and $\partial L/\partial \sigma = 0$. An important role in these equations is played by the Fisher score function

$$\psi(x) = -\frac{f_0'(x)}{f_0(x)} = -\frac{d}{dx} \log f_0(x). \quad (8.2)$$

Between ψ and the score function $J_0(u) = -[f_0 Q_0(u)]'$, there is an important relation:

$$J_0(u) = \psi[Q_0(u)]. \quad (8.3)$$

The maximum likelihood estimators $\hat{\mu}$ and $\hat{\sigma}$ are the solutions of

$$\int_0^1 \psi\left(\frac{\tilde{Q}(u) - \hat{\mu}}{\hat{\sigma}}\right) du = 0, \quad (8.4)$$

$$\int_0^1 \psi\left(\frac{\tilde{Q}(u) - \hat{\mu}}{\hat{\sigma}}\right) \{\tilde{Q}(u) - \hat{\mu}\} du = \hat{\sigma}.$$

Under the symmetry assumption one seeks robust estimators of location; various standard estimators may be heuristically motivated by approximating (8.4) in suitable ways.

M-estimators are defined by introducing a window

$$w(x) = \frac{1}{x} \psi(x).$$

Then (8.4) may be written

$$\int_0^1 w\left(\frac{\tilde{Q}(u) - \hat{\mu}}{\hat{\sigma}}\right) \{\tilde{Q}(u) - \hat{\mu}\} du = 0.$$

To estimate $\hat{\mu}$, consider the limit of an iterative sequence $\hat{\mu}^{(n)}$ defined by

$$\hat{\mu}^{(n+1)} = \frac{\int_0^1 w[(\tilde{Q}(u) - \hat{\mu}^{(n)})/\hat{\sigma}]\tilde{Q}(u)\,du}{\int_0^1 w[(\tilde{Q}(u) - \hat{\mu}^{(n)})/\hat{\sigma}]\,du}. \tag{8.5}$$

The estimator $\hat{\mu}^{(n)}$ is an M-estimator. For $w(x)$ one could choose a function which corresponds to Student's t-distribution with m degrees of freedom (Parzen, 1979b):

$$w(x) = \frac{m+1}{m} \frac{1}{1 + (1/m)x^2}.$$

Various widely used choices of $w(x)$ are described in Hogg (1979). The most widely used choice for $w(x)$ may be Tukey's bisquare window

$$w(x) = \left[1 - \left(\frac{x}{c}\right)^2\right]_+^2,$$

where c is a suitable constant, often chosen as 6. The choice of m or c is crucial; it should reflect one's beliefs about how long are the tails of $F_0(x)$.

L-estimators are linear combinations of order statistics which can be written in terms of the quantile function $\tilde{Q}(u)$ as follows, for a suitable weight function $W_\mu(u)$:

$$\hat{\mu} = \frac{\int_0^1 \tilde{Q}(u) W_\mu(u)\,du}{\int_0^1 W_\mu(u)\,du}. \tag{8.6}$$

If the model $Q = \mu + \sigma Q_0$ is assumed to hold, with a symmetric Q_0, and one chooses

$$W_\mu(u) = \psi'[Q_0(u)] \tag{8.7}$$

then $\hat{\mu}$ is an asymptotically efficient estimator of μ. A rigorous derivation of (8.6) and of (8.10) can be obtained from eqn (9.3). A heuristic derivation of (8.6) from (8.4) is obtained by writing

$$\psi\left(\frac{\tilde{Q}(u) - \hat{\mu}}{\hat{\sigma}}\right) = \psi[Q_0(u)] + \psi'[Q_0(u)]\left\{\frac{\tilde{Q}(u) - \hat{\mu}}{\hat{\sigma}} - Q_0(u)\right\}. \tag{8.8}$$

Since $\psi[Q_0(u)]$ and $Q_0(u)$ are odd functions, and $\psi'[Q_0(u)]$ is even, the estimation equations for $\hat{\mu}$ are

$$0 = \int_0^1 \psi\left(\frac{\tilde{Q}(u) - \hat{\mu}}{\hat{\sigma}}\right) du = \int_0^1 \psi'[Q_0(u)]\left\{\frac{\tilde{Q}(u) - \hat{\mu}}{\hat{\sigma}}\right\} du. \tag{8.9}$$

From (8.9) one obtains the estimator defined by (8.6) and (8.7).

2 DATA MODELING USING QUANTILE AND DENSITY-QUANTILE FUNCTIONS

An estimator of σ which is asymptotically efficient for the model $Q = \mu + \sigma Q_0$, when F_0 is a symmetric distribution, is

$$\hat{\sigma} = \frac{\int_0^1 \tilde{Q}(u) W_\sigma(u)\,du}{\int_0^1 Q_0(u) W_\sigma(u)\,du}, \tag{8.10}$$

where

$$W_\sigma(u) = J_0(u) + Q_0(u) W_\mu(u). \tag{8.11}$$

It is often the case that $W_\sigma(u)$ is approximately equal to $J_0(u)$ (times a constant such as 1 or 2). This helps explain why the following definition works.

Score deviations and minimum residual score deviation estimators

It is a remarkable fact that one can define a universal (and robust) measure of deviation of a sample:

$$\tilde{\sigma}_0 = \int_0^1 f_0 Q_0(u) \tilde{q}(u)\,du. \tag{8.12}$$

Assuming that $f_0 Q_0(u) \tilde{Q}(u) = 0$ for $u = 0, 1$, we can write $\tilde{\sigma}_0$ in the form

$$\tilde{\sigma}_0 = \int_0^1 J_0(u) \tilde{Q}(u)\,du, \tag{8.13}$$

which we call a sample *score deviation*. To calculate it one has to specify a score function $J_0(u)$. Note that $\tilde{\sigma}_0$ estimates a population quantity defined by

$$\sigma_0 = \int_0^1 J_0(u) Q(u)\,du, \tag{8.14}$$

which we call a score deviation.

Robust estimators of location called R-estimators can be interpreted as minimum "residual score deviation" estimators. More precisely suppose one estimates the location parameter μ by an estimator $\hat{\mu}$ whose residuals $\tilde{Q}(u) - \hat{\mu}$ have smallest deviation:

$$\int_0^1 J_0(u)\{\tilde{Q}(u) - \mu\}\,du \text{ is minimized};$$

it can be shown that this is precisely the definition of R-estimators.

M-estimators can also be motivated from this point of view; to avoid specifying $J_0(u) = \psi_0[Q_0(u)]$ in (8.13) one replaces it by $\psi[(\tilde{Q}_0(u)-\mu)/\sigma]$ and the criterion to estimate μ is to minimize

$$\int_0^1 \psi\left(\frac{\tilde{Q}(u)-\mu}{\sigma}\right)\{\tilde{Q}(u)-\mu\}\,du = \int_0^1 w\left(\frac{\tilde{Q}(u)-\mu}{\sigma}\right)\{\tilde{Q}(u)-\mu\}^2\,du,$$

whose solution might be sought as the limit of sequences $\hat{\mu}^{(n)}$ of the form of (8.5).

The fact that R- and M-estimators of μ can be formulated as minimum residual score deviation estimators seems to explain why these methods can be extended to estimation of regression coefficients. However L-estimators do not have a natural generalization to regression. Further R- and M-estimators yield asymptotically equivalent results when their $J_0(u)$ and ψ functions satisfy (8.3).

9. Data summary by a few values of the sample quantile function

To form an estimated quantile function $\hat{Q}(u)$, the simplest approach is to first attempt to fit a parametric family of the form

$$Q(u) = \mu + \sigma Q_0(u), \tag{9.1}$$

where $Q_0(u)$ is a specified quantile function; μ and σ are called location and scale parameters since $F(x) = F_0[(x-\mu)/\sigma]$. One seeks to form estimators $\hat{\mu}$ and $\hat{\sigma}$ which are asymptotically efficient under the hypothesis that the true quantile function satisfies (9.1).

Some of the *aims* for which the quantile function approach to statistical data analysis may provide rigorous, yet simple, *methods* are as follows:

(1) To provide approximately efficient estimators of μ and σ under the hypothesis $H_0: Q(u) = \mu + \sigma Q_0(u)$.
(2) To perform quick goodness-of-fit tests of H_0, and/or to find re-expressions (transformations) of the data which satisfy H_0.
(3) To perform rigorous goodness-of-fit tests to identify quantile functions Q_0 for which the data satisfies H_0, and/or to estimate the density-quantile function.

Estimation of μ and σ

Efficient and tractable estimators of μ and σ which are linear functionals in $Q(u)$ can be found using the theory of regression analysis of continuous parameter time series developed by Parzen (1961). The asymptotic distribution theory of $\tilde{Q}(u)$ permits us to write approximately

2 DATA MODELING USING QUANTILE AND DENSITY-QUANTILE FUNCTIONS

$$\sqrt{nf Q(u)}\{\tilde{Q}(u) - Q(u)\} = B(u).$$

Under H_0,

$$Q(u) = \mu + \sigma Q_0(u), \qquad fQ(u) = \frac{1}{\sigma} f_0 Q_0(u).$$

Consequently, defining $\sigma_B = \sigma/\sqrt{n}$,

$$f_0 Q_0(u) \tilde{Q}(u) = \mu f_0 Q_0(u) + \sigma f_0 Q_0(u) Q_0(u) + \sigma_B B(u). \tag{9.2}$$

The parameter σ_B is linearly related to σ, but it is here treated as a free parameter. In terms of the inner product of the RKHS of the Brownian Bridge covariance kernel one may express the minimum variance unbiased estimators $\hat{\mu}$ and $\hat{\sigma}$ given $\tilde{Q}(u)$ for $0 \leq u \leq 1$, or $p \leq u \leq q$, or $u = u_1, \ldots, u_k$, as follows

$$\begin{pmatrix} \hat{\mu} \\ \hat{\sigma} \end{pmatrix} = \mathrm{Inf}^{-1} \begin{pmatrix} \langle f_0 Q_0, (f_0 Q_0) \tilde{Q} \rangle \\ \langle (f_0 Q_0) Q_0, (f_0 Q_0) \tilde{Q} \rangle \end{pmatrix}, \tag{9.3}$$

Table 4. Order Statistics Optimal for Location Parameter Estimation by Seven Order Statistics

	Distribution			
Spacing	Normal	Cauchy	Logistic	Extreme value
0.01				✓
0.02	✓			✓
0.0625				✓
0.125	✓	✓	✓	✓
0.1875				
0.25		✓	✓	✓
0.3125	✓			
0.375		✓	✓	
0.4375				✓
0.5	✓	✓	✓	
0.5625				
0.625		✓	✓	
0.6875	✓			✓
0.75		✓	✓	
0.8125				
0.875	✓	✓	✓	
0.9375				
0.98	✓			
0.99				

Table 5. Order Statistics Optimal for Scale Parameter Estimation by Seven Order Statistics

				Distribution				
Spacing	Exponential or Weibull	Pareto $v = 0.5$	Pareto $v = 1$	Pareto $v = 2$	Pareto $v = 3$	Logistic	Normal	Log-normal
0.01							✓	
0.02						✓	✓	
0.0625		✓				✓		✓
0.125		✓✓						✓
0.1875		✓	✓			✓		✓
0.25		✓	✓					
0.3125	✓		✓	✓				
0.375			✓	✓	✓	✓		✓
0.4375			✓	✓	✓			
0.5								
0.5625	✓			✓	✓	✓		
0.625				✓	✓			
0.6875				✓✓	✓			
0.75	✓		✓					
0.8125	✓✓					✓		
0.875	✓✓					✓		✓
0.9375	✓✓					✓	✓	✓
0.98	✓✓						✓✓	✓
0.99	✓							✓

where Inf is the information matrix,

$$\text{Inf} = \begin{pmatrix} \langle f_0 Q_0, f_0 Q_0 \rangle & \langle f_0 Q_0, (f_0 Q_0) Q_0 \rangle \\ \langle Q_0 (f_0 Q_0), f_0 Q_0 \rangle & \langle (f_0 Q_0) Q_0, (f_0 Q_0) Q_0 \rangle \end{pmatrix}. \quad (9.4)$$

The variance–covariance matrix of $\hat{\mu}, \hat{\sigma}$ equals $\sigma_B^2 \{\text{Inf}\}^{-1}$.

It should be emphasized that the foregoing expressions are not valid if $f_0 Q_0(u)$ and $[f_0 Q_0(u)] Q_0(u)$ do not belong to the RKHS, which can happen in the case of the index set $0 \leq u \leq 1$. Failure to belong to the RKHS seems to be equivalent to the optimal parameter estimator involving a few extreme value order statistics, which implies that the estimators are not asymptotically normal.

If one could accomplish these aims using a "few" (say, seven) selected order statistics, then one could regard these "few" order statistics as an efficient summary of the entire sample of size n. If large samples (as well as small samples) could be effectively represented by a small number of order statistics, then every data set could be published and each reader could easily do "hands on" statistical data analysis.

The problem of choosing order statistics for the estimation of location and scale parameters has an extensive literature. The density-quantile approach has been investigated by Eubank (1979) in his Ph.D. thesis. By using location and scale estimators based on only 7 quantile values for a specified Q_0, one can identify 19 quantile values which are the union of these 7 values over a large number of familiar choices of Q_0. The proposed 19 number universal data summary consists of the median $Q(0.5)$; the $j/16$ percentiles $Q(j/16)$, $Q[(8+j)/16]$ for $j = 7, 6, 5, 4, 3, 2, 1$; and the 0.01 and 0.02 percentiles $\tilde{Q}(0.01)$, $\tilde{Q}(0.02)$, $\tilde{Q}(0.98)$, $\tilde{Q}(0.99)$. Tables 4 and 5 are from Eubank's thesis, and show which of these order statistics are used to estimate location and scale parameters of familiar probability laws.

References

Akaike, H. (1973). Information theory and an extension of the maximum likelihood principle. In *2nd International Symposium on Information Theory* (B. H. Petrov and F. Csaki, eds), 267–281. Akademiai Kiado, Budapest.

Csorgo, M. and Revesz, P. (1978). Strong approximations of the quantile process, *Annals of Statistics*, **6**, 882–894.

Effron, B. (1979). Bootstrap methods: another look at the jackknife, *Annals of Statistics*, **7**, 1–26.

Eubank, R. L. (1979). A Density-Quantile Function Approach to Choosing Order Statistics for the Estimation of Location and Scale Parameters. Technical Report A-10, Institute of Statistics, Texas A & M University.

Hogg, R. V. (1979). Statistical robustness: one view of its applications today, *American Statistician*, **33**, 108–115.

Ord, J. K. and Patil, G. K. (1975). Statistical modelling: an alternative view. In *Statistical Distributions in Scientific Work* (G. P. Patil, S. Kotz and J. K. Ord, eds), Vol. 2, 1–9. Reidel, Boston.

Parzen, E. (1961). Regression analysis of continuous parameter time series. In *Proceedings of the 4th Berkeley Symposium on Mathematical Statistics and Probability*, Vol. 1, 469–489. University of California Press, Berkeley.

Parzen, E. (1979a). Nonparametric statistical data modeling, *Journal of the American Statistical Association*, 105–131.

Parzen, E. (1979b). A density-quantile function perspective on robust estimation. In *Robustness in Statistics* (R. Lanner and G. Wilkinson, eds), 237–258. Academic Press, New York.

Tukey, J. W. (1965). Which part of the sample contains the information, *Proceedings of the National Academy of Sciences*, **53**, 127–134.

Tukey, J. W. (1977). *Exploratory Data Analysis*. Addison-Wesley, Reading, Massachusetts.

Zacks, S. (1971). *The Theory of Statistical Inference*. Wiley, New York.

3

STOCHASTIC PROCESSES IN THE THEORY OF EPIDEMICS

J. Gani

Abstract

After a brief historical introduction, some discrete time stochastic models for epidemics are outlined. Three continuous time stochastic models are then presented and their practical uses discussed. Finally, simple spatial models for epidemics, mainly in one dimension are considered. The conclusion is drawn that stochastic models have proved invaluable in illuminating epidemic research.

1. Historical introduction

Graunt (1662) was probably the earliest author to treat epidemics quantitatively in his *Bills of Mortality*. In the second chapter of his book, entitled "General Observations upon the Casualties", Graunt analyzed various causes of death, and foreshadowed the theory of competing risks. He pointed out, for example, that of 229 250 deaths recorded over 20 years in the London parishes, only 5500 (or 1 in every 41) were due to such "notorious diseases" as apoplexy, gout, jaundice and sciatica, whereas the far greater number of 16 000 (or 1 in every 14) were caused by the dreaded plague.

Approximately a century later Bernoulli (1760) presented the first mathematical paper on epidemic theory to the Académie Royale des Sciences, Paris. This analyzed the effects of smallpox on mortality, and supported variolation as a protective measure against the disease. Bernoulli was able to formulate a deterministic model in continuous time t for the number of susceptibles $x(t)$ who had not contracted smallpox in a cohort of size $\xi(t) \geqslant x(t)$, $t \geqslant 0$. Solving a differential equation for $x(t)$, he obtained its relation to

$\xi(t)$, and from this derived the number of individuals $\zeta(t)$ who would have remained alive in a state without smallpox. Comparing $\xi(t)$, based on Halley's (1693) life tables, with the calculated values of $\zeta(t)$, Bernoulli showed that if smallpox were eliminated, the cohort size of survivors aged 25 years would be increased by just over one-seventh.

The mid-nineteenth and early twentieth centuries saw the beginnings of empirical work on epidemic data; Farr and Brownlee carried out statistical analyses of curves describing the growth and decline of infectives, and deaths, during the course of an epidemic (Fine, 1979). Farr (1840), assuming that the second or third ratios of successive weekly mortality figures were constants, had been able to predict subsequent mortalities with some measure of success. Brownlee (1907) showed that epidemic data could be fitted by Pearson curves or a mixture of normals; he later explained his observations with the hypothesis of decreasing infectivity.

But no mechanism for the spread of infection had yet been broadly accepted as a suitable model for epidemic processes. It was Hamer (1906) who first put forward the assumption which is basic to all modern epidemic theory: namely that the incidence of infectives in a population following an incubation period depends on the contact rate between susceptibles and infectives just before it. In its deterministic form, with the discrete time t measured in units of incubation periods, this "mass action" principle states that

$$y_{t+1} = \beta x_t y_t, \qquad t = 0, 1, 2, \ldots, \tag{1.1}$$

where x_t, y_t are the numbers of susceptibles and infectives, respectively, at time t, and β is the infection parameter.

The continuous time version

$$\frac{dy}{dt} = \beta xy, \qquad t \geq 0, \tag{1.2}$$

where $x(t)$, $y(t)$ are, respectively, susceptibles and infectives at time t, provides a formulation of the simple epidemic in a closed population for which $x + y = N$. For the general epidemic, in which there is an additional class $z(t)$ of removals, with $x + y + z = N$, this is modified by a further factor $-\gamma y$ on the right-hand side of (1.2), where this denotes the rate of removals of infectives by recovery or death, γ being the removal parameter. The most commonly used stochastic models are based on a modification of these "mass action" results, as we shall see in the following sections.

2. Early and discrete time stochastic models

The first stochastic treatment of epidemic phenomena was published by McKendrick (1926). He was concerned essentially with mean numbers v_x of individuals who had already experienced x attacks of a disease; the probability of a further attack in time δt was then $f_{t,x}\delta t + o(\delta t)$ so that

$$\frac{dv_x}{dt} = f_{t,x-1}v_{x-1} - f_{t,x}v_x. \tag{2.1}$$

An explicit solution of this equation is possible if $f_{t,x} = f_x = b + cx$; when $c/b \to 0$, this leads to a Poisson result for v_x.

McKendrick also studied the two-dimensional case where

$$\frac{dv_{x,y}}{dt} = f_{x-1,y}v_{x-1,y} - f_{x,y}v_{x,y} + g_{x,y-1}v_{x,y-1} - g_{x,y}v_{x,y}, \tag{2.2}$$

and $v_{x,y}$ could, for example, represent the number of infected individuals at the point (x, y), the probabilities $f_{x,y}\delta t + o(\delta t)$, $g_{x,y}\delta t + o(\delta t)$ being related to horizontal and vertical movement on the lattice.

His stochastic approach was not followed up for another 20 years. Meanwhile Reed and Frost in their lectures as early as 1928 (Abbey, 1952) and Greenwood (1931) developed stochastic chain binomial models of infection in discrete time. These models were not recognized as Markovian until several decades later, when Gani (1969) and Gani and Jerwood (1971) formulated their theory in terms of simple and bivariate Markov chains.

Let us consider a closed population of $x_0 = N$ susceptibles, into which infection is introduced by $y_0 \geq 1$ infectives, and for which the chance of contact with a susceptible is $0 < p = 1 - q < 1$. In the Greenwood model, the probability of infection is assumed not to depend on the number of infectives present, but only on the presence of infection, so that if X_t, X_{t+1} represent the random numbers of susceptibles at times $t, t+1$, respectively ($t = 0, 1, 2, \ldots$), then

$$\Pr\{X_{t+1} = x_{t+1} | X_t = x_t\} = \binom{x_t}{x_{t+1}} p^{x_t - x_{t+1}} q^{x_{t+1}}, \tag{2.3}$$

where the number of infectives at time $t + 1$ is $y_{t+1} = x_t - x_{t+1}$.

The transition matrix for the states $X_t = 0, 1, \ldots, N$ is therefore given by the triangular matrix

$$M = \begin{bmatrix} 1 & & & & \\ p & q & & & \\ p^2 & 2pq & q^2 & & \\ \cdots & \cdots & \cdots & \cdots \\ p^N & Np^{N-1}q & \cdots & q^N \end{bmatrix} \tag{2.4}$$

in which all elements above the main diagonal are zero.

If we now write $Q' = [1, q, \ldots, q^N]$ and $P = M - \operatorname{diag} Q$ we can readily find the generating function of the time T to termination of the epidemic, when $X_T = X_{T-1}$ or the number of infectives is $Y_T = 0$; this is given by

$$[0, 0, \ldots, 0, 1](I - \theta P)^{-1} \theta Q, \quad 0 \leqslant \theta \leqslant 1. \tag{2.5}$$

It is also possible to write the joint probability generating function of T and X_T and thus to obtain the total size of the epidemic $N - X_T$.

For the Reed–Frost model, in which the number of infectives Y_t is now taken into account, the chance of infection at time t being $(1 - q^{y_t})$, we have

$$\Pr\{X_{t+1} = x_{t+1}, Y_{t+1} = y_{t+1} \mid X_t = x_t, Y_t = y_t\} = \binom{x_t}{x_{t+1}} (1 - q^{y_t})^{y_{t+1}} q^{y_t x_{t+1}}. \tag{2.6}$$

$\{X_t, Y_t\}$ now form a bivariate Markov chain; it is possible to write the transition matrix for its $(N+1)^2$ states, where each of X_t, Y_t can take the values $0, 1, \ldots, N$. Apart from the somewhat more complicated form of the transition matrix, calculations for this case are analogous to those for the Greenwood model, and probability generating functions for T and X_T can be derived with equal facility.

One of the main interests of these chain binomial models lies in the estimation of the probability p of infectious contact. Wilson et al. (1939) used them to analyze household data for measles epidemics in Providence, Rhode Island during 1929–34. These yielded values of $\hat{p}$ in the neighborhood of 0.65 to 0.8 for both Greenwood and Reed–Frost formulations, depending on the particular model and the method of analysis employed. The same methods have also been usefully applied to diseases such as the common cold, influenza, chicken pox and mumps. It is interesting to note that variations have been found in the value of p as between different households, and also between different cities. Nevertheless, these stochastic methods can usefully predict the likely spread and progress of an epidemic in a population, when it is subdivided into smaller household units.

3. Continuous time stochastic models

The stochastic equivalent of the simple epidemic (1.2) in continuous time $t \geqslant 0$, in a closed population of N individuals, is given by the model in which

$$\Pr\{X(t+\delta t) = x - 1 \mid X(t) = x\} = \beta xy \, \delta t + o(\delta t) = \beta x(N - x) \, \delta t + o(\delta t) \tag{3.1}$$

denote the infinitesimal probabilities of the relevant Markov chain in continuous time. This has been treated in detail by Bailey (1957), who refined his results in two subsequent papers (Bailey, 1963, 1968).

The forward Kolmogorov equations for the probabilities

$$p_i(t) = \Pr\{X(t) = i | X(0) = N-1\}, \quad i = 0, 1, \ldots, N-1,$$

given that there is one initial infective $Y(0) = 1$ in the total population N, are

$$\frac{dp_i}{dt} = \beta(i+1)(N-i)p_{i+1} - \beta i(N-i+1)p_i, \quad 0 \leq i \leq N-2 \tag{3.2}$$

$$\frac{dp_{N-1}}{dt} = -\beta(N-1)p_{N-1}$$

subject to $p_{N-1}(0) = 1$. If $\mathbf{P}'(t) = [p_0(t), \ldots, p_{N-1}(t)]$, with the initial $\mathbf{P}'(0) = [0, 0, \ldots, 0, 1]$, then we can write these as

$$\frac{d\mathbf{P}}{dt} = Q\mathbf{P}, \tag{3.3}$$

where the elements of the matrix Q are given by $q_{ii} = -\beta i(N-i+1)$ ($0 \leq i \leq N-1$) in the diagonal, $q_{i,i+1} = \beta(i+1)(N-i)$ ($0 \leq i \leq N-2$) in the super-diagonal, and zero elsewhere.

The problem is solved in principle, since it is known that

$$\mathbf{P}(t) = e^{Qt}\mathbf{P}(0), \tag{3.4}$$

but the explicit expressions for the probabilities $p_i(t)$ are not entirely straightforward in their details.

The duration T of such an epidemic is a period of importance; it can be obtained as the sum of $N-1$ exponential random variables with parameters $\beta i(n-i)$ ($1 \leq i \leq N-1$). Kendall (1957) has derived asymptotic results for the associated random variable

$$W = NT - 2\ln N - 1, \tag{3.5}$$

whose moment generating function is found to be $\{\Gamma(1-\theta)\}^2$ with probability density function

$$f(w) = 2\exp(-w)K_0[2\exp(-\tfrac{1}{2}w)], \quad -\infty < w < \infty, \tag{3.6}$$

where K_0 is the modified Bessel function of the second kind.

The most commonly used continuous time model, however, is that for the general stochastic epidemic, equivalent to Kermack and McKendrick's (1927) deterministic model for susceptibles $x(t)$ and infectives $y(t)$. These satisfy the differential equations

$$\frac{dx}{dt} = -\beta xy, \quad \frac{dy}{dt} = \beta xy - \gamma y \tag{3.7}$$

with the removals given by $z = N - x - y$, and initial values set at $x_0 = N - a$, $y_0 = a$, $z_0 = 0$. The Markovian model is now bivariate with infinitesimal probabilities

$$\Pr\{X(t+\delta t) = x-1, Y(t+\delta t) = y+1 | X(t) = x, Y(t) = y\} = \beta xy\, \delta t + o(\delta t),$$
$$\Pr\{X(t+\delta t) = x, Y(t+\delta t) = y-1 | X(t) = x, Y(t) = y\} = \gamma y\, \delta t + o(\delta t). \quad (3.8)$$

For this epidemic process, Bartlett (1949) initiated researches into the Kolmogorov equations for the probabilities

$$p_{ij}(t) = \Pr\{X(t) = i, Y(t) = j | X(0) = N-a, Y(0) = a\}.$$

These are given by

$$\frac{dp_{ij}}{dt} = \beta(i+1)(j-1)p_{i+1,j-1} - j(i\beta + \gamma)p_{ij} + \gamma(j+1)p_{i,j+1},$$

$$0 \leq i \leq N-a, 0 \leq j \leq N, 0 \leq i+j \leq N \quad (3.9)$$

$$\frac{dp_{N-a,a}}{dt} = -a[\beta(N-a) + \gamma]p_{N-a,a}$$

subject to $p_{N-a,a}(0) = 1$. The probability generating function $P(z, w, t) = \sum_{i,j} p_{ij}(t) z^i w^j$ satisfies the partial differential equation

$$\frac{\partial P}{\partial t} = \beta(w^2 - wz)\frac{\partial^2 P}{\partial z \partial w} + \gamma(1-w)\frac{\partial P}{\partial w} \quad (3.10)$$

with $P(z, w, 0) = z^{N-a} w^a$.

Attempts by Gani (1965, 1967) and Siskind (1965) to solve the eqns (3.10) or (3.9) explicitly have been only partially successful, though in principle, one can write

$$\mathbf{P}'(t) = [p_{00}(t), \ldots, p_{0N}(t); p_{10}(t), \ldots, p_{1N-1}(t); \ldots; p_{N-a,0}(t), \ldots, p_{N-a,a}(t)]$$

and again express the Kolmogorov equations in the form (3.3), with solution (3.4) where once more $\mathbf{P}'(0) = [0, 0, \ldots, 0, 1]$.

As with the chain binomial models, this formulation for the general epidemic, with some additional complexities, allows the estimation of the parameters β and γ (or $\rho = \gamma/\beta$); these are estimated, for example, in Bailey and Thomas (1971). Here some smallpox data from Nigeria is analyzed to give values $\hat{\beta} = 0.00088 \pm 0.00025$, $\hat{\gamma} = 0.091 \pm 0.031$. But this model can be used not only to obtain quantitative predictions of infectives and removals; with modifications to take account of the Poisson immigration of susceptibles and infectives it is also able to describe the qualitative behavior of epidemics, as Bartlett's (1957, 1960a) studies of recurrent measles epidemics has demonstrated. Bartlett was able to show, among other results, that oscillations in the number of measles infectives tended to persist when the size of the community was above a certain critical level $N \sim 250\,000$, while the infection tended to fade out if the community size was smaller. Data from both the United Kingdom and the United States have broadly confirmed this finding.

If we allow a Poisson immigration factor with parameter μ for susceptibles and ε for infectives, the process now has a probability generating function $P(z, w, t)$ satisfying the partial differential equation

$$\frac{\partial P}{\partial t} = \beta(w^2 - wz)\frac{\partial^2 P}{\partial z \partial w} + \gamma(1-w)\frac{\partial P}{\partial w} + \mu(z-1)P + \varepsilon(z-1)P. \quad (3.11)$$

We can study the stable distribution obtained as $t \to \infty$, and find the mean recurrence time $\bar{R}$ to extinction of infectives in terms of the initial number of susceptibles x_0 and infectives y_0. If, as in Bartlett (1960b), we set $x_0/y_0 = 34.1$, a value which holds approximately for the city of London, and take $\gamma = \frac{1}{2}$ week, then we find that

$$\bar{R} \sim \frac{\{2\pi(x_0+y_0)\}^{1/2}}{\gamma y_0} \exp\left\{\left(y_0 + \frac{x_0}{y_0}\right)^2 \bigg/ 2(x_0+y_0)\right\} \quad (3.12)$$

varies between roughly a year for $y_0 = 130$, $x_0 = 4433$, to 24 years for $y_0 = 400$, $x_0 = 13\,640$.

4. Spatial models

A realistic representation of the spread of an epidemic must take into account the spatial or topographical component. It is surprising that after McKendrick's (1926) initial attack on the spatial problem, little advance on it was made for the next 30 years. In their books, Bartlett (1960b) and Bailey (1975) have given brief accounts of the initial work carried out in this area.

To provide a simple example of a spatial model, let us consider the deterministic case of the two-dimensional continuous spread of individuals with density $\sigma(\mathbf{r})$ at the vector position $\mathbf{r}$. At time $t \geq 0$, the local density of susceptibles and infectives will be

$$x(\mathbf{r}, t)\sigma(\mathbf{r})r\,dr\,d\theta, \qquad y(\mathbf{r}, t)\sigma(\mathbf{r})r\,dr\,d\theta,$$

where $x(\mathbf{r}, t)$, $y(\mathbf{r}, t)$, $z(\mathbf{r}, t)$ are now proportions of susceptibles, infectives and removals, respectively, such that

$$x(\mathbf{r}, t) + y(\mathbf{r}, t) + z(\mathbf{r}, t) = 1.$$

The analogues of the Kermack–McKendrick equations (3.7) at the position

$\mathbf{r}_0$ are given by

$$\frac{\partial x}{\partial t} = -x\bar{y} = -x \iint y(\mathbf{r},t)\sigma(\mathbf{r})f(|\mathbf{r}-\mathbf{r}_0|)r\,dr\,d\theta \qquad (4.1)$$

$$\frac{\partial y}{\partial t} = x\bar{y} - \frac{\rho}{\sigma}y.$$

Here, the time scale is now altered so as to make the infection parameter $\beta = 1$, the double integral is taken over all values of $\mathbf{r}$, $f(|\mathbf{r}-\mathbf{r}_0|)$ accounts for the effect of distance from $\mathbf{r}_0$, and ρ is taken to have the dimensions of a density, like σ.

These equations are ordinarily very difficult to solve, but some progress can be made in the simplest case of a population distributed along the line $\mathbf{r} = (r, \theta = 0)$ if we follow Kendall's (1965) treatment. In this, it is assumed that $\bar{y}$ is a short-range symmetrical average, so that it can be written as the diffusion approximation

$$\bar{y} = y + k\frac{\partial^2 y}{\partial r^2}, \qquad k > 0. \qquad (4.2)$$

Equations (4.1) now reduce to

$$\frac{\partial x}{\partial t} = -x\left(y + k\frac{\partial^2 y}{\partial r^2}\right), \qquad \frac{\partial y}{\partial t} = x\left(y + k\frac{\partial^2 y}{\partial r^2}\right) - \lambda^{-1}y \qquad (4.3)$$

with $\lambda^{-1} = \rho/\sigma$.

Kendall was interested in traveling waves along the line and so set

$$x(r, t) = x(r - ct), \qquad y(r, t) = y(r - ct)$$

in which $c > 0$ is the velocity of the wave, the form of the wave being maintained as t increases. Writing $S = r - ct$, we see that ahead of the wave

$$S = \infty, \qquad x = 1, \qquad \frac{dx}{dS} = 0, \qquad y = \frac{dy}{dS} = 0, \qquad (4.4)$$

whereas behind the wave

$$S = -\infty, \qquad x < 1, \qquad \frac{dx}{dS} = 0, \qquad y = \frac{dy}{dS} = 0. \qquad (4.5)$$

The differential eqns (4.3) are transformed to

$$-c\frac{dx}{dS} = -x\left(y + k\frac{d^2y}{dS^2}\right), \qquad -c\frac{dy}{dS} = x\left(y + k\frac{d^2y}{dS^2}\right) - \lambda^{-1}y. \qquad (4.6)$$

from which we can readily obtain

$$c\ln x - k\frac{dy}{dS} = c\lambda(x + y) + A.$$

But from (4.4) we can see that the constant $A = -c\lambda$, so that Kendall finally obtains

$$\frac{dy}{dx} = \frac{(\lambda^{-1}\ln x - x + 1) - y}{(1 + k\lambda^{-2}c^{-2})y - (\lambda^{-1}\ln x - x + 1)} = \frac{U(x) - y}{by - U(x)}. \quad (4.7)$$

We need to look for solution curves terminating for $S = \infty$ at $x = 1, y = 0$, and starting for $S = -\infty$ at $x = \alpha < 1, y = 0$. Also, since $dx/dS = dy/dS = 0$ at both ends, dy/dx must be indeterminate, so that $U(\alpha) = U(1) = 0$. For $\alpha < 1$, $U(\alpha) = 0$ only if $\lambda > 1$, that is if $\sigma > \rho$; this leads to Kendall's threshold theorem for the population density which must exceed the critical value ρ if waves are to develop.

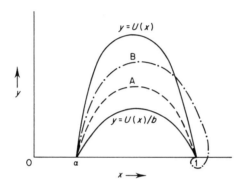

Fig. 1. Solution curves.

Let us consider the solution curve when this condition is satisfied. This is unique and lies between $y = U(x)$ and $y = U(x)/b$. Two possibilities A or B (Fig. 1) may occur depending on whether $c \geqslant c_{\min}$ or $c < c_{\min}$, where

$$c_{\min} = 2\left[k\left(1 - \frac{\rho}{\sigma}\right)\right]^{1/2}. \quad (4.8)$$

If $c < c_{\min}$, the solution curve B will spiral into the half-plane $y < 0$, which is not permissible; thus no waves with speed lower than $c_{\min}$ are possible. If, however, $c \geqslant c_{\min}$ then curve A will always hit $x = 1$ from the half-plane $y \geqslant 0$, so that waves of any speed greater than $c_{\min}$ are possible.

Stochastic equivalents of this, and other more complex spatial models have been reviewed by Mollison (1977), who has developed them in the broader framework of Markov contact processes. For the case of infection along the line, suppose that the distance of the furthest infective from the starting point 0 of the epidemic is

$$R_t = \max_i r(i), \quad (4.9)$$

where $i \in P(t)$ the set of infectives. Then the front velocity is defined as $v_t = R_t/t$, while the expected instantaneous velocity is $(\partial/\partial t)E(R_t)$, if this expectation exists and is differentiable.

For the case of the general stochastic epidemic (with an infinite pool of susceptibles), if $W(r,t)$ is the expected value at time t of the proportion of infectives $Y(r,t)$ at position r, then for this birth-death Markov contact process, $W(r,t)$ satisfies the differential equation

$$\frac{\partial W}{\partial t} = \bar{W} - \rho W. \qquad (4.10)$$

The expected instantaneous velocity $(\partial/\partial t)E(R_t)$ is finite, and satisfies

$$\frac{\partial}{\partial t} E(R_t) \leq c'_v, \qquad (4.11)$$

provided the migration distribution has an exponentially bounded tail. Daniels (1977) has found an expression for the asymptotic expectation velocity c'_v, and Faddy (1977) has outlined a discrete approximation method for the general stochastic epidemic. Once again the stochastic model characterizes the qualitative behavior of wave formation in epidemics spreading in one or two dimensions. Considerable work is continuing in the area of spatial stochastic processes (see, for example, Renshaw, 1979).

5. The role of stochastic processes in epidemiology

For some epidemic problems, deterministic models provide an adequate description of the underlying process of infection. Such is the case in Kermack and McKendrick's (1927) continuous time model of the general epidemic, for which they have derived their well-known "threshold theorem". This qualitative result states that an epidemic will develop only if the initial number of susceptibles x_0 in a closed population N is larger than the relative removal rate $\rho = \gamma/\beta$; if $x_0 = \rho + \nu$ ($\nu > 0$), then as $t \to \infty$, $x_\infty = \rho - \nu$ at the end of the epidemic. Satisfactory quantitative predictions are also possible on the basis of discrete time deterministic models, such as Spicer's (1979) influenza model for England and Wales.

But deterministic models fail to account for the more subtle characteristics of epidemics, in particular their variability, which is dependent on the random nature of infectious phenomena. Without chain binomial theory, the spread of infection through households could hardly be predicted accurately on a quantitative basis. Continuous time Markovian models for the simple and general epidemics have allowed the estimation of relevant infection and removal parameters and the numerical simulation of infective counts. At the qualitative level, the mechanism of recurrence in measles epidemics could not have been elucidated without Bartlett's (1957, 1960b) stochastic modeling and simulation. Lastly, the velocity of infective waves in one and two dimensions have been much clarified in the framework of spatial stochastic models, and readily simulated in discrete time on lattices.

Where variation needs to be accounted for in epidemics, stochastic models are essential. They have explained qualitatively some basic characteristics of infectious processes, and provided quantitative predictions and realistic simulations of these. In such ways have stochastic processes illuminated many of the problems of epidemic research. It seems likely that they will remain an invaluable and increasingly used tool in epidemiology.

References

Abbey, H. (1952). An examination of the Reed–Frost theory of epidemics. *Human Biology*, **24**, 201–233.
Bailey, N. T. J. (1957). *The Mathematical Theory of Epidemics*, 1st edn. Griffin, London.
Bailey, N. T. J. (1963). The simple stochastic epidemic: a complete solution in terms of known functions, *Biometrika*, **50**, 235–240.
Bailey, N. T. J. (1968). A perturbation approximation to the simple stochastic epidemic in a large population, *Biometrika*, **55**, 199–209.
Bailey, N. T. J. (1975). *The Mathematical Theory of Infectious Diseases and its Applications*. Griffin, London.
Bailey, N. T. J. and Thomas, A. S. (1971). The estimation of parameters from population data on the general stochastic epidemic, *Theor. Pop. Biol.* **2**, 253–270.
Bartlett, M. S. (1949). Some evolutionary stochastic processes, *Journal of the Royal Statistical Society* B, **11**, 211–229.
Bartlett, M. S. (1957). Measles periodicity and community size, *Journal of the Royal Statistical Society* A, **120**, 48–70.
Bartlett, M. S. (1960a). The critical community size for measles in the United States, *Journal of the Royal Statistical Society* A, **123**, 37–44.
Bartlett, M. S. (1960b). *Stochastic Population Models in Ecology and Epidemiology*. Methuen, London.
Bernoulli, D. (1760). Essai d'une nouvelle analyse de la mortalité causée par la petite vérole, et des avantages de l'inoculation pour la prévenir, *Mém. Math. Phys. Acad. Roy. Sci.*, Paris, 1–45. (In *Histoire de l'Académie Royale des Sciences*, 1766).
Brownlee, J. (1907). Statistical studies in immunity. The theory of an epidemic, *Proceedings of the Royal Society, Edinburgh*, **26**, 484–521.
Daniels, H. E. (1977). The advancing wave in a spatial birth process, *Journal of Applied Probability*, **14**, 689–701.
Faddy, M. J. (1977). Stochastic compartmental models as approximations to more general stochastic systems with the general stochastic epidemic as an example, *Advances in Applied Probability*, **9**, 448–461.
Farr, W. (1840). Progress in epidemics. *Second Report of the Registrar-General of England and Wales*, 16–20.

Fine, P. E. M. (1979). John Brownlee and the measurement of infectiousness: an historical study in epidemic theory, *Journal of the Royal Statistical Society* A, **142**, 347–362.
Gani, J. (1965). On a partial differential equation of epidemic theory, I, *Biometrika*, **52**, 617–622.
Gani, J. (1967). On the general stochastic epidemic. In *Proceedings of the 5th Berkeley Symposium on Mathematical Statistics and Probability*, Vol. 4, 271–279. University of California Press, Berkeley.
Gani, J. (1969). A chain binomial study of inoculation in epidemics, *Bulletin of the International Statistical Institute*, **43**, 203–204.
Gani, J. and Jerwood, D. (1971). Markov chain methods in chain binomial epidemic models, *Biometrics*, 591–603.
Graunt, J. (1662). *Natural and Political Observations Made upon the Bills of Mortality.* London (reprinted by The Johns Hopkins Press, Baltimore, 1939).
Greenwood, M. (1931). On the statistical measure of infectiousness, *Journal of Hygiene, Cambridge*, **31**, 336–351.
Halley, E. (1693). An estimate of the mortality of mankind, drawn from curious tables of the births and funerals at the city of Breslaw; with an attempt to ascertain the price of annuities upon lives, *Philosophical Transactions of the Royal Society, London*, **17**, 596–610.
Hamer, W. H. (1906). Epidemic disease in England—the evidence of variability and of persistency of type, *The Lancet*, **ii**, 733–739.
Kendall, D. G. (1957). La propagation d'une épidémie on d'un bruit dans une population limitée, *Publ. Inst. Statist. Univ. Paris*, **6**, 307–311.
Kendall, D. G. (1965). Mathematical models of the spread of infection. In *Mathematics and Computer Science in Biology and Medicine*, 213–225. HMSO, London.
Kermack, W. O. and McKendrick, A. G. (1927). Contributions to the mathematical theory of epidemics, I, *Proceedings of the Royal Society* A, **115**, 700–721.
McKendrick, A. G. (1926). Applications of mathematics to medical problems, *Proceedings of the Edinburgh Mathematical Society* **14**, 98–130.
Mollison, D. (1977). Spatial contact models for ecological and epidemic spread, *Journal of the Royal Statistical Society* B, **39**, 283–326.
Renshaw, E. (1979). Waveforms and velocities for non-nearest-neighbour contact distributions, *Journal of Applied Probability*, **16**, 1–11.
Siskind, V. (1965). A solution of the general stochastic epidemic, *Biometrika*, **52**, 613–616.
Spicer, C. C. (1979). The mathematical modelling of influenza epidemics, *British Medical Bulletin*, **35**, 23–28.
Wilson, E. B., Bennett, C., Allen, M. and Worcester, J. (1939). Measles and scarlet fever in Providence, R.I., 1929–1939 with respect to age and size of family. *Proceedings of the American Philosophical Society*, **80**, 357–476.

4
THE BAYESIAN APPROACH TO STATISTICS

Dennis V. Lindley

Abstract

This paper discusses several topics that arise in applying Bayesian ideas to inference problems. The Bayesian paradigm is first described as an appreciation of the world through probability: probability being expressed in terms of gambles. Various justifications for this view are outlined. The role of models in the specification of probabilities is considered; together with related problems of the size and complexity of the model, robustness and goodness of fit. Some attempt is made to clarify the concept of conditioning in probability statements. The role of the second argument in a probability function is emphasized again in discussion of the likelihood principle. The relationship between the probability specification and real-world experiences is explored and a suggestion is made that zero probabilities are, in a sense, unreasonable. It is pointed out that it is unrealistic to think of probability as necessarily being defined over a σ-field. The paper concludes with some remarks on two common objections to the Bayesian view.

1. Introduction

The Bayesian approach to statistics is a complete, logical framework for the discussion and solution of problems of inference and of non-competitive

This research was supported by the Air Force Office of Scientific Research (AFSC), USAF, under Grant AFOSR-77-3179 while the author was a Visiting Professor at the Operations Research Center, University of California, Berkeley. Reproduction in whole or in part is permitted for any purpose of the US Government.

decision-making. It has many facets, from axiomatic foundations of probability to sophisticated technical manipulations needed to solve practical problems. The present paper deals only with inference and is devoted to some questions of Bayesian *philosophy*. We explore some of the points that arise when the mathematics is related to the real world, avoiding both the completely logical questions within the mathematics and the technicalities of real-world situations, and concentrating on our ways of thinking within the paradigm. A wider view is given in Lindley (1971) with some additional comments in Lindley (1978).

In Section 2 the Bayesian paradigm is defined: essentially as a probabilistic view of the scientific world. The meaning of probability is explained. Practical difficulties in so appreciating the whole of our environment are formidable and we naturally need to view parts in isolation: how this can be done is considered in Section 3. The important role of conditioning, especially on new information, is the topic of Section 4. All statistics, and much of science, use models to describe phenomena. The meaning and role of models in Bayesian statistics is discussed in Section 5, and continued in Section 6 where the problem of the fit of a model is considered. In Section 7 the relationship of a Bayesian view of the world to the reality of that world is investigated and a suggested strengthening of the paradigm to exclude certain undesirable behaviour is introduced. In Section 8 replies are offered to two of the most oft-repeated criticisms of the Bayesian attitude.

There is no pretense to completeness in the range of topics discussed: they are points that seem to me to be of some importance and on which I may have a little to add. My debt to de Finetti is considerable. Often, when I feel I have something new to say, I realize that all I am really doing is appreciating the significance of some of his writings. Sometimes, unfortunately, I may have misinterpreted some of his views. The potential reader of this paper might better spend his time with de Finetti (1972, 1974/75).

2. The Bayesian paradigm

The scientist's appreciation of the world—and I use the possessive to remind us that poets, musicians, artists and others see the world differently—is of a collection of *quantities*, of things that can be described by numbers. For a scientist to understand and manipulate things, he must measure them, or at least think of them as things that, indirectly or directly, might be measurable. For example, human preferences are studied by measuring them in terms of quantities called utilities. Without this quantification the scientist cannot proceed: with it, he has at his disposal the full force of logic and the mathematical argument. The success of the scientific approach depends, in part, on how well this quantification encapsulates the situation under study. At the moment it does rather well in mechanics; less well in psychological studies of preference.

These quantities are of two types: those whose numerical value is known to the scientist, and those which are unknown. The members of the first group are familiar to us as known, real numbers. The unknown quantities are more mysterious. We shall refer to them as *random quantities* (RQ), though the term, uncertain quantity, is often used. A random quantity will be denoted by a capital letter, say X; its numerical value by the corresponding lower-case letter, x. Collections of random quantities will not be distinguished notationally from single quantities. When X becomes known then x will be replaced by the revealed number. Thus X may be the breadth of the desk on which I write: x its numerical value in feet. When the desk is measured it may be found that $x = 4$ and X, as a random quantity, becomes known to be 4 ft. An important subclass of random quantities refers to events, where $X = 1$ if the event is true, and $X = 0$ if it is false.

At any point in time the scientist contemplates a set of quantities some of which are known, some unknown. The values of those in the first set describe part of what he knows at the moment. He will also know about some logical relations that exist between the quantities, both known and unknown: for example, the area of the desk is the product of its length and breadth. This knowledge of logic will not appear explicitly in our notation but its presence must not be forgotten; a point we will return to in Section 3. The set of *known* quantities will be denoted by H and their values by h. Similarly the set of random quantities will be written, X. Whilst it is easy to describe H, namely as h, more elaborate description is needed for random quantities. Although a random quantity is, by its nature, unknown, it is never completely unknown, in the sense that the scientist knows nothing about it. In the example of X, the breadth of the desk, in feet, one would think 2 or 3 much more reasonable values than 500, or 0.002; and -5 is illogical. It is possible to study random quantities such as X, the desk's breadth, almost as though they were merely letters that the mathematician can so powerfully and profitably manipulate, and forget that X is the representation of a real thing. This forgetfulness leads to difficulties in describing the uncertainty which are substantially diminished as soon as the reality behind the letter is remembered. The scientist's problem is to describe this partial knowledge that he has of the random quantities that interest him.

The Bayesian position is that this uncertainty, or partial knowledge, is to be described in terms of probability. Thus the value of 3 ft is more probable for my desk than is that of 500 ft. In general, with X the set of random quantities and H the set of known ones; the Bayesian, scientific description of the world is a probability distribution of X, given H: this is written $p(X|H)$

and read "the probability of X given H". It is most important to recognize that when it is claimed that the description is in terms of probability it is not merely meant that the description is by means of a number lying between 0 and 1. Much more is intended: namely that different uncertainties are related by the rules of probability. We remind the reader of these rules in the case of random events.

Convexity

$$0 \leqslant p(X|H) \leqslant 1 \quad \text{and} \quad p(H|H) = 1.$$

Addition

If X_1 and X_2 are exclusive, meaning the event, X_1 and X_2 both true, is logically impossible, then

$$p(X_1 \text{ or } X_2|H) = p(X_1|H) + p(X_2|H).$$

Multiplication

$$p(X_1 \text{ and } X_2|H) = p(X_1|H)p(X_2|X_1 \text{ and } H).$$

Thus a geneticist who knows that a cross-fertilized plant must be red, pink or white and expresses his uncertainty by saying that the probability of red is 0.4, of pink 0.2, is committed to a probability of 0.6 that the plant will be coloured. The addition and multiplication rules describe how uncertainties combine, or cohere, and a set of uncertainty statements obeying the probability rules is often said to be *coherent*. The Bayesian position is that uncertainties should cohere in this technical sense. Before proceeding further, we digress in the next paragraph to discuss a few points of notation.

In general X and H will be multidimensional and continuous so that $p(X|H)$ will be a probability density for X with respect to some dominating measure. To avoid technical problems that do not concern this paper we shall often think in terms of discrete random quantities and use summation in place of the more general integration with respect to the dominating measure. The reader may object that since H is known, as h, it is unnecessary to consider $p(X|H)$ and that $p(X|h)$ would suffice. There are two reasons for preferring the extended form. First, we often need to consider several possible values of H. Thus suppose a scientist is designing a spacecraft to visit Venus. He is uncertain about the temperature Y on the surface and needs to contemplate the performance X of an instrument in various temperature conditions: thus he contemplates $p(X|Y)$ for all temperatures. This point will recur again when we consider in Section 4 more precisely what is meant by "given H" in the probability of X, given H.

The second reason for using H instead of h is technical. If, as in the Venus example, Y is continuous, it transpires that a satisfactory definition of conditional probability is only possible for a random quantity Y and not for

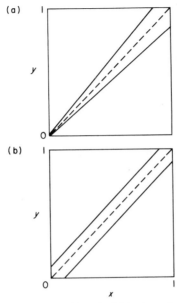

Fig. 1. A point (X, Y) uniformly distributed within the unit square. (a) $Z = X/Y$ near to 1, (b) $W = X - Y$ near to 0.

a single measurement where the quantity is unspecified. This is usually known as Borel's paradox. To illustrate, let a point, described by its random coordinates X and Y, be uniformly distributed in the unit square. It is meaningless to consider the uncertainty in X given that the point lies on the diagonal through the origin. For if the diagonal is described by $Z = X/Y = 1$ and measurements are of Z, the situation is quite different from when it is described by $W = X - Y = 0$. This can be seen by considering values of Z near 1 and W near 0: Z near 1 (Fig. 1a) means X is more likely to be near 1 than near 0; whereas W near 0 (Fig. 1b) means all values are equally likely.

Still on notation, we shall sometimes write $p_{X,H}(\cdot|\cdot)$ instead of the shorter $p(X|H)$ to emphasize that we are dealing with functions, in this case of two variables, X and H. For example, $p_{X,H}(2|3)$, in the case of one random quantity, X and one known, H, means the probability that $x = 2$, given that $h = 3$: here the variables have taken the numerical values 2 and 3.

To return to the Bayesian paradigm, asserting that uncertainties about quantities should cohere according to the rules of probability, we need to ask why they should combine this way. There are two answers, *axiomatic* and *pragmatic*. In outline, the axiomatic approach proceeds by searching for simple, self-obvious truths about uncertainty, taking these as axioms and developing a mathematical system of theorems and proofs from them. As an example of such a truth, consider three uncertain events A, B and C for which a scientist thinks that A is more likely than B and B more likely than C. Then it seems self-evident that he should think A more likely than C. From a set of such axioms it is possible to prove that the relationship "more likely than" just mentioned corresponds to an ordering by means of probability, obeying the three rules described above. The first person to attempt such a development successfully was Ramsey (1964)—the original publication was 1931. The first detailed discussion was by Savage (1954). A good, modern exposition is by DeGroot (1970). Many scientific studies use the axiomatic approach; Euclidean geometry and Newtonian mechanics are good examples, because if the axioms encapsulate the real-world situation accurately the theorems will prove useful guides to action. Modern statistical inference outside the Bayesian paradigm lacks such an axiomatic structure.

The above approach is essentially within the field of *pure* mathematics; in order to use it, to pass to *applied* mathematics, it is necessary to have an operational meaning for the terms that occur in the mathematics. What do we mean when we say that the probability that $X = 2$, given $h = 3$, is 0.4? How can we operate with such a number? The concept of probability has been much discussed with frequentist, logical and subjective interpretations. Here the subjectivist view will be adopted. Consider an urn containing a white and b black balls and a gamble in which a prize will be won if, on drawing a ball at random from the urn, it is white. Contrast this with a gamble in which the same prize will be won if a random event A is true. If the scientist is indifferent between the two gambles we say the probability for A is $a/(a+b)$. Any increase (decrease) in a will make the urn gamble more (less) attractive. Notice that this interpretation can be tested, to see if the scientist is so indifferent: we do not suggest it is necessarily the best test. Ways of testing, using coherence, have been considered by Lindley et al. (1979).

(Notice that in the interpretation the phrase "drawing a ball at random" has been used. This means that were a prize to be given if a particular ball was drawn, it would not matter which ball was selected to decide the prize. Loosely, all balls are equally likely. The interpretation is not circular.)

The gambling interpretation leads to the *pragmatic* justification for the Bayesian view: simply that it works. The axioms are prescriptions for one's attitude to uncertainty and they lead to the general, Bayesian prescription or recipe. Let X denote the random quantities of interest: let H denote the known quantities. Then the *Bayesian recipe* is to calculate $p(X|H)$, the calculation to proceed using the calculus of probabilities: nothing more, nothing less. My claim is that the recipe works: try it for yourselves and see.

An alternative approach which is, in some ways, intermediate between the axiomatic and pragmatic ones, is due to de Finetti (1974/75). Suppose, on considering an uncertain event E, a scientist describes his uncertainty by a number x, knowing that if E subsequently turns out to be true he will receive a penalty score $(x-1)^2$, whereas if it is false he will receive x^2. Then from the single principle, that if the scientist gives $x, y, \ldots$ for $E, F, \ldots$, the scores being additive, he will not choose values that are such that other values exist for which, whatever is true, the score is reduced, de Finetti shows that the given values must be probabilities. This approach has the merits of the axiomatic approach in being developed from few principles and yet also, in the score, gives an operational interpretation for the value.

According to Popper, a requirement of a good theory is that it should be possible to make predictions which can be tested in practice. If the test succeeds, the theory is enhanced; any failure, and the theory is damaged. Our experience with the probability calculus is extensive and many predictions can be made and tested: there is no case known to me where they fail. This is not to say that the predictions can be evaluated in all cases: on the contrary, there are many situations where the technical difficulties are formidable and unsolved. But this does not invalidate the coherent approach any more than did the initial failure to solve the three-body problem invalidate Newtonian mechanics. In this essay I want to concentrate on the recipe inherent in the Bayesian approach: on the constructive methodology produced by the description of random quantities in terms of probability.

3. Large and small worlds

In principle X and H could embrace all that we do not and do know: in practice we confine ourselves to relatively few quantities. It is therefore important to see how contraction can occur. In the case of the random quantities the situation is straightforward. Let X be decomposed into (X_1, X_2) and suppose that only X_1 is of interest. Then by the addition rule

$$p(X_1|H) = \sum_{X_2} p(X_1, X_2|H)$$

to obtain the *marginal* distribution of X_1. This summation or, in general, integration is basic to the Bayesian recipe and is essentially the addition rule. An important example of its use is in the elimination of nuisance parameters to concentrate on the parameters of interest. Thus in a large agricultural experiment with many varieties and treatments, we can investigate a pair of varieties by integration over treatments and plot constants that appear in the design. Lest this remark appear trivial, let us point out an important practical application. Consider an experiment to compare several treatments and let X_i denote the uncertain yield from treatment i. Then, after experimentation, one has $p(X_1, X_2, \ldots, X_k|H)$ for data H, but one may be interested in comparing a treatment that seems to have done rather well with one that has done badly.

From the probability just cited one can calculate $p(X_i - X_j | H)$ to effect the comparison. Such techniques are not available in sampling-theory statistics and a body of knowledge has been developed, called multiple comparisons, to deal with the problem. In the Bayesian view, no special techniques are needed and the solution is a straightforward use of the addition rule.

If we need to pass from X_1 back to X we can do this by

$$p(X_1, X_2 | H) = p(X_1 | H) p(X_2 | X_1, H)$$

combining the original marginal for X_1 with $p(X_2 | X_1, H)$. This is the multiplication rule.

Comparable changes in the known quantities are not so transparent. Let H be decomposed into H_1, H_2. Then

$$p(X | H_1) = \sum_{H_2} p(X | H_1, H_2) p(H_2 | H_1),$$

which involves the previously unconsidered $p(H_2 | H_1)$, a number that may be hard to assess since it involves forgetting the known quantities, H_2. There is one case where the reduction is simple, namely when

$$p(X | H_1, H_2) = p(X | H_1)$$

or X is independent of H_2, given H_1. The assumption of independence is often made because it obviates the need to consider $p(H_2 | H_1)$, or even H_2, at all. Often it is made without the scientist consciously thinking about it, as when he ignores information in a colleague's paper, H_2, when considering his own. One of the most attractive features of the Bayesian recipe is its ability to put all information together coherently, rather than consider it piecemeal, so that all assumptions of independence need to be considered carefully.

An expansion from H_1 to $H = (H_1, H_2)$ usually arises in studying the most important tool of science, experimentation or observation. The basic idea behind observation is that a previously unknown, or random, quantity becomes known. In our term, it passes from the first argument of a probability function to the second argument. Let (X_1, X_2) be random quantities and suppose X_2 is observed, then

$$p(X_1 | X_2, H) = \frac{p(X_1, X_2 | H)}{\sum_{X_1} p(X_1, X_2 | H)}$$

providing a rule for the passage from the original uncertainty of both X_1 and X_2 to that of X_1 given X_2. Notice that no new uncertainties are required, only the margin for X_2. This result is more often used by applying the product rule to the numerator to give

$$p(X_1|X_2, H) = \frac{p(X_2|X_1, H)p(X_1|H)}{\sum_{X_1} p(X_1, X_2|H)}$$

This is Bayes' rule. It is perhaps the most important rule for human understanding of the real world, for it tells us exactly how we should incorporate the observation of X_2 into our scientific appreciation of that world. Our fuller notation, with X_1 as the variable, expresses it more clearly

$$p_{X_1}(\cdot|x_2, H) \propto p_{X_2}(x_2|\cdot, H)p_{X_1}(\cdot|H), \tag{3.1}$$

the omitted constant of proportionality not involving the variable. The fundamental role played by Bayes' result gives the recipe its name: it might be more sensible to term it the *coherent recipe* since Bayes was only the first of many who played a role in its development.

We have seen how X and H can be changed. Let us now consider how extensive to make the sets of random and known quantities: how big should we make our scientific world? In an idealization of scientific behaviour, it would be chosen to embrace everything, both known and unknown. Such a scientist would never need to rethink his understanding of the world, for as soon as X_2, part of $X = (X_1, X_2)$, becomes known he could calculate $p(X_1|X_2, H)$ by Bayes' rule. Such a situation is rarely practicable though it does reasonably arise in one problem of some importance, namely the study of a finite population, where $X = X^{(n)} = (X_1, X_2, \ldots, X_n)$, X_i being the uncertain quantity for the ith member of the population. Observation of a sample from the population changes $p(X^{(n)}|H)$ into $p(X_{m+1}, \ldots, X_n | X^{(m)}, H)$ where $X^{(m)}$ is the sample.

That situation is unusual in that initial specification of all probabilities is a practical possibility. Usually the scientist prefers to specify such probabilities as seen to him at the time to be relevant and only consider others when the need arises. Let us take a simple case of this and explore it in some detail, since that will reveal other important features of the probabilistic understanding of the world.

Suppose that a scientist considers two events A and B. He attaches probabilities to these, given H, $\alpha = p(A|H)$ and $\beta = p(B|H)$. The only constraints imposed by the coherence requirements are that both α and β lie in the unit interval. Furthermore there is no obligation on the scientist to consider other events logically derivable from A and B: for example, the union of A and B, $A \cup B$. It is usually stated that the events under consideration must form a σ-field, closed under countable unions and intersections. In a sensible description of reality this is not so, the events may have any structure that the scientist finds convenient; and all that the coherence requirements demand is that those probabilities he has *actually* assessed (and not those he *might* have assessed) obey the rules of the calculus of probabilities. Suppose next that the scientist does some logic and discovers that A and B are exclusive, $A \cap B = \emptyset$. Then H has changed because of this logical consideration, to H^1 say, and he could now assess $p(A|H^1)$ and $p(B|H^1)$ as α^1 and β^1, respectively. There is no exact connection between the original assessments, α and β, and the revised values, α^1 and β^1, obtained after the logical calculation. All that can be said is that, assuming $\alpha \leq \beta$ without loss of generality, $0 \leq \alpha^1 \leq \alpha$ and $(\beta-\alpha)/(1-\alpha) \leq \beta^1 \leq \beta$. (An explanation for these inequalities follows in a few lines.)

If, in addition to A and B, the scientist considers the union and assigns probability $\gamma = p(A \cup B|H)$, he has then implicitly assessed $p(A \cap B|H)$ to be $(\alpha+\beta-\gamma)$. Consequently if he now performs the logic and finds that A and B are exclusive, he can update his original probabilities, rescaling them to add to one. Hence $\alpha^1 = (\gamma-\beta)/(1-\alpha-\beta+\gamma)$ and $\beta^1 = (\gamma-\alpha)/(1-\alpha-\beta+\gamma)$. (Since $\max(\alpha,\beta) = \beta \leq \gamma \leq \alpha+\beta$, these equalities provide the inequalities stated at the end of the previous paragraph.) So we see that it is possible for logical calculations to change probabilities according to rules of the calculus, as in this paragraph, or to change them arbitrarily subject only to some inequality constraints, as in the previous paragraph. The role of logical considerations in formulating uncertain opinions is therefore a subtle one that perhaps deserves more consideration than has heretofore been given. De Finetti (1974/75) has investigated the condition under which a set of probabilities, such as α, β and γ above, do constitute a complete set in the sense that all other events, logically derivable from those considered, have their probabilities implied by the rules of the calculus.

An example of incomplete logic that is often quoted concerns the decimal expansion of π, the ratio of the circumference to the diameter of a circle. For most of us the probability that the millionth digit is 7 is 0.1, despite the fact that were we to do the logic—equivalent to finding $A \cap B = \emptyset$ above—we would know the value of the millionth digit. This is a clear example of how H can change by purely logical considerations. Bayes' theorem principally describes how H changes by empirical considerations. It is a practical advantage of the coherent scheme that it can embrace situations in which the logic is incomplete so that, for example, not all the events in the σ-field need be prescribed.

4. Conditioning

It should be clear that probability is a function of *two* arguments, X and H. Most treatments of probability are based on probability as a measure. Probability as a function of X is a measure; but as a function of H, it is not.

4 THE BAYESIAN APPROACH TO STATISTICS

The measure-theory accounts diminish the role of the second argument and reinforce this by utilizing independence conditions so substantially that manipulations using the second argument are scarcely needed. We have already seen how independence can simplify the reduction of H and we shall see later how it can be used with great advantage in a Bayesian treatment of models in Section 5. Nevertheless independence needs most careful consideration. For example, independence is itself a conditional statement: we cannot say A is independent of B, but only given H; for the independence may fail if H changes. An excellent discussion is given by Dawid (1979).

The role of the second argument in probability is most clearly revealed in Bayes' theorem,

$$p_{X_1}(\cdot | x_2, H) \propto p_{X_2}(x_2 | \cdot, H) p_{X_1}(\cdot | H).$$

These two measures for X_1 are connected by the function $p_{X_2}(x_2 | \cdot, H)$ of X_1 in the other argument. Whilst $p_{X_2}(\cdot | x_1, H)$ is a measure, $p_{X_2}(x_2 | \cdot, H)$ in the theorem is not. It is called a *likelihood* function; in contrast to a probability (or measure) function. Its importance is considerable because the only feature of the random quantity X_2, that has become known to be x_2, used in Bayes' result is the likelihood function of X_1 for that value x_2 (and not for other values of X_2). This is natural since X_2 has passed from being random to being known and the previously possible, but now impossible, values are irrelevant. We say the likelihood function of X_1, given x_2, is *sufficient*. Much of modern statistics denies this sufficiency and requires more from X_2 than the likelihood function: and it is this denial that constitutes the main difference between the Bayesian and other paradigms of statistics. Notice that in accord with the earlier discussion of Borel's paradox, it is not enough to know x_2, we must know which RQ it was that was seen to have the value x_2. X_2 may not be forgotten entirely.

In considering the role of the two arguments in a probability statement attention needs directing towards the word "given" in the phrase "probability of X, given H". A scientist, at a point in time, will know H and consider X. In that case it is sensible to say "given H" as an abbreviation for "given H is known". But we have seen that he might evaluate $p(X_1 | X_2, H)$, still at the point in time where X_2 is unknown. In that case "given X_2" does not mean X_2 is known but rather "were X_2 to be known": the subjunctive tense is the relevant one. The gambling interpretation of a probability has been explained above, and a gamble on X_1, were X_2 to be true, is interpreted as a gamble on both X_1 and X_2 that only pays if both are true: indeed, a form of this identification constitutes one of the axioms in the theoretical development of the coherent paradigm. It is sometimes referred to as the axiom of *called-off bets*, meaning that the gamble is called off if X_2 (as an event) is false. Hence in contemplating $p(X_1 | X_2, H)$ the scientist is thinking of his attitude to X_1 *were*

X_2 true. There is a distinction between the contemplation of X_2 and the actual experience of X_2. There is no reason why, when X_2 is actually realized and the subjunctive becomes unnecessary, the scientist should not express his uncertainty about X_1 differently from the original $p(X_1|X_2, H)$. He must still adhere to the original gamble but his attitude to new gambles might be different. All of us experience the situation in which the experience of something changes our view. Thus I might evaluate my probability of something were it to rain tomorrow, but change it when I experience the rain. This does not invalidate the necessity for the original judgements to cohere, but gives a license to change when contemplation becomes reality. What happens here is that the scientist would say that it is not merely the rain that I have experienced but other, previously unconsidered events, like wet clothes, that I had not contemplated. In other words, that my original collection of random quantities was too small. This agrees with our previous observation that it is not enough to consider "given x_2" but that we need to consider "given X_2" for all realizations x_2 of X_2. We must always be alert to a possible useful enlargement of the sets of quantities being considered.

Another point that arises here is the difference between "F is true" and "F is known to be true". Really, E given F means the contemplation of E "were F known to be true". This is clear because the probabilistic description of the world is of an incomplete world with lots of true and false events whose truth or falsity is not known to the scientist, so that the probability must incorporate this lack of knowledge. An interesting example of the need for the distinction arises in legal applications. Suppose that a crime has been committed and the forensic evidence establishes the fact that the criminal has a certain property, A. We later learn that an individual, Smith, has A: what is the probability that Smith is the criminal? We need to calculate the probability that "Smith is guilty" given "we know that Smith has A". As soon as one expresses it this way, one realizes that it might be pertinent to ask how it is that we know Smith has A. One possibility is that when the forensic evidence was described to a police officer he reported that he knew a man with A, his name was Smith. Another possibility is that the police searched amongst their files, or amongst possible suspects, until they found someone with A, who chanced to be Smith. Detailed calculations show that the two methods of acquiring the knowledge that Smith has A lead to different numerical values for the required probability of Smith's guilt.

The point is related to Borel's paradox where we conditioned not just on x but on a random quantity X that has taken the value x. We cannot condition on Smith's having A but on the observation of some random quantity that takes that value. The quantity in a police search is different from that of a policeman's recollection. It is also related to the question of whether a quantity is controlled or not. Thus an experiment might be performed in which a quantity X varies from unit to unit. Such an experiment may give us valuable information for a new unit in which X is uncertain, but little, or no, information for a new unit controlled, or made, to have a prescribed value for X. To appreciate what happens when control is exercised one has to experiment with control. Free and controlled variations differ not in x, but in the X realized to be x.

5. Models

The interpretation of probability in terms of a gamble is only meaningful if the gamble can be settled; that is, if the status of the random quantity can be changed to a known quantity either by observation or by logic. The same remark applies to de Finetti's device of a scoring rule. This is no real restriction since it is idle to talk about things that cannot affect observable quantities. It is idle to discuss the random event that Shakespeare wrote the plays attributed to him unless the event has realizable effects—and doubtless the tourist industry in Stratford feels that it has. We like to talk about unobservables because they can influence observables about which gambles can be settled. A more important reason is that the introduction of unobservables can simplify our probability considerations. Let us see how this can happen.

The specification of $p(X|H)$ is difficult if only because the dimensionalities of X and H are both large, and we have seen that there are advantages in considering large sets. We therefore seek ways of simplifying the specification. We can always write

$$p(X|H) = \sum_\theta p(X|\theta, H) p(\theta|H)$$

for any quantity θ. Suppose now that X is independent of H, given θ; then our assessment of the uncertainty about X can be changed to one conditional on θ but for which H is irrelevant, and to another about θ. In particular if θ can have low dimensionality, the specifications required may be simpler. We say $p(X|\theta)$ is then a (probability) *model* and refer to θ as a *parameter* of the model. A model is self-contained in the sense that it is uninfluenced by the known quantities in H and, more importantly, by a change in H. If X is decomposed into (X_1, X_2) we may write

$$p(X_1, X_2|H) = \sum_\theta p(X_1|\theta) p(X_2|\theta, X_1) p(\theta|H).$$

If the further assumption is made that θ known not merely suppresses H but also an addition of H to $H \cup X_1$, we may write

$$p(X_1, X_2|H) = \sum_\theta p(X_1|\theta) p(X_2|\theta) p(\theta|H) \qquad (5.1)$$

resulting in yet further simplification since X_1 and X_2 necessarily have smaller dimensionalities than X. Most of the models used in statistics have this further property that make not only X and H independent given θ, but also X_1 and X_2 independent given θ. Notice that "given" is again to be interpreted here in the subjunctive form since the parameter is rarely an observable quantity that can be made known. Rather its role is to stand between X and H to give them, and the components of X, independence properties. A further simplification that is often made is to suppose in (5.1) that the distributions of X_1 and X_2 given θ are functionally the same: in our fuller notation $p_{X_1}(\cdot|\theta) = p_{X_2}(\cdot|\theta) = f(\cdot|\theta)$ say. Generalizing from 2 to n random quantities we have the common statistical model with

$$p(X^{(n)}|H) = \sum_\theta \prod_{i=1}^n f(X_i|\theta)p(\theta|H). \tag{5.2}$$

It is usual to make inferences about the parameter

$$p(\theta|X^{(n)}, H) \propto \prod_{i=1}^n f(X_i|\theta)p(\theta|H)$$

but it is often more realistic to make inferences in terms of observables. Thus we may write

$$p(X_n|X^{(n-1)}, H) = \sum_\theta f(X_n|\theta)p(\theta|X^{(n-1)}, H) \tag{5.3}$$

in virtue of the independence of X_n and $X^{(n-1)}$, given θ. The role of the model and of the parameter is to simplify the calculations and to isolate extraneous factors in H from the rest of the system. In particular, the specification of the probabilities given H are reduced from $p(X|H)$ to $p(\theta|H)$. Even the latter may be too complicated and we may prefer to model that through

$$p(\theta|H) = \sum_\phi p(\theta|\phi)p(\phi|H)$$

and *hyperparameters* ϕ; with θ independent of H, given ϕ. The process may be repeated to model the hyperparameters; and so on. The device is essentially one to simplify our calculations and descriptions.

A sequence $X^{(n)}$ satisfying (5.2) has the property of *exchangeability* given H: that is, the probability is invariant under permutation of the suffixes $1, 2, \ldots, n$, as is easily seen because of the common function f. (Not all exchangeable sequences have the representation (5.2), but the more commonly-used ones do.) This exchangeability implies a particular form of connection between X_n and $X^{(n-1)}$ as spelt out in (5.3). It is an important part of our study of uncertainty to recognize exchangeability and to see, for a random quantity Y, which sequences $X^{(n-1)}$ are exchangeable with Y. This establishes a relation between a random Y and data $x^{(n-1)}$ of known quantities. The point has been considered by Lindley and Novick (1981).

6. Model fit

Although the use of models in expressing uncertainty about quantities is widespread, it has long been recognized that no model is a totally adequate description of one's uncertainty, and that, at best, it is a good approximation. To appreciate this consider the following argument where M is an event of high probability; specifically $p(\bar{M}|H) = \varepsilon$ with ε small. Then

$$p(X|H) = p(X|M,H)p(M|H) + p(X|\bar{M},H)p(\bar{M}|H)$$

or

$$p(X|H) - p(X|M,H) = \{p(X|H,\bar{M}) - p(X|H,M\}p(\bar{M}|H).$$

Since the difference of probabilities in braces cannot exceed 2 in modulus,

$$|p(X|H) - p(X|M,H)| \leq 2\varepsilon.$$

This result enables us to condition on an event, M, of high probability without making much difference to the final result: or we may condition on an event that is nearly true. If M is the event that a model, such as (5.2), is true, then we may evaluate $p(X|H)$ as if (5.2) obtained provided we have high probability for the model. If $X = (X_1, X_2)$ and X_2 becomes known, so that H changes to H, X_2, the same argument will persist provided $p(\bar{M}|H, X_2)$ remains small. However, the observed value x_2 may suggest that M is false and then the calculations will be seriously affected by the supposition of M. Thus the Bayesian paradigm supports the principle of using a simple model until the data suggests it might be false.

One way of proceeding is to think of a model with parameter θ as part of a wider model with parameter (θ, α) that reduces to the earlier model when $\alpha = 0$. If $p(\alpha = 0|H)$ is large we may ignore α, but if x_2 reduces this probability seriously then it may be advisable to consider values of α other than zero. The description of this is not too difficult. One would start, in the usual continuous case, with a density for α centred at zero with small spread and then update this by Bayes' theorem to evaluate $p(\alpha|X_2, H)$.

It therefore pays to make the model as large as possible. Since an important reason for introducing the model is to simplify the calculations, these two considerations pull in opposite directions and some compromise is essential. There are cases where the technical manipulations are not so hard that very large models may be contemplated. As an example, consider a situation where one wishes to make inferences about the relationship between two, one-dimensional quantities Y and X partly expressed through $E(Y|X)$. We may model this through $E(Y|X,\theta)$ plus other features of the distributions of Y, given X. Over a finite range of X the expectation may be described by an expression

$$E(Y|X,\theta) = \sum_{i=0}^{\infty} f_i(X)\theta_i,$$

where $f_i(X)$ is a polynomial of degree i orthogonal to the other polynomials and $\theta = \{\theta_1, \theta_2, \ldots\}$. To perform the Bayesian analysis it is necessary to describe the uncertainty about this polynomial expressed through uncertainty about the θ_is. Our general scientific experience teaches us that polynomials of rather low degrees suffice, so that the distribution for θ_i with large i would concentrate around zero. The necessary calculations have been performed by Young (1977). There we have a model which, at least as far as the expectation is concerned, could hardly be any larger and no difficulties over its approximations should arise.

Notice that the adequacy of the model can be investigated within the Bayesian paradigm by the calculation of $p(\bar{M}|H, X_2)$, or, in parametric form, using $p(\alpha|H, X_2)$. In practice, the first suggestion that M is inadequate to explain the data will arise through informal considerations, such as plotting the residuals, but a formal investigation requires the scientist to think about $\bar{M}$; that is, about alternatives to M. We return to this point later in the section.

This combination of informal suggestion of an alternative to the model and its subsequent, more precise analysis has been discussed because it has been suggested, most recently by Box (1980), that the Bayesian paradigm is not adequate to deal with model criticism and that devices outside the coherent system are necessary. One such device is to test the hypothesis that the model is M by calculating $p(X_2|H, M)$ which express the uncertainty, given M, of the random quantity subsequently observed and using this to obtain $p[E(x_2)|H, M]$ where $E(x_2)$ is a set of values of X_2 "more extreme than" the observed x_2. (A common example of "more extreme than" is the tail of the distribution of a univariate X_2 beyond x_2.) There is a conceptual connection between the choice of what constitutes more extreme values and the alternative models described by α. Whilst the latter consideration is coherent the former is not. One can easily see this by noting that in order to evaluate $p[E(x_2)|H, M]$ more information about X_2 has to be provided beyond x_2 as a known value of X_2, which is all that the coherent paradigm requires. In particular, values that could fall in $E(x_2)$ have to be contemplated. As Jeffreys remarks, what might have happened [to X_2] but didn't, hardly seems relevant once $X_2 = x_2$ has been observed. The sufficiency of the likelihood function is being denied.

These considerations are also relevant to the closely related question of robustness of an inference procedure wherein we ask how far the final uncertainty $p(X_1|X_2, H)$ is affected by the choice of model. Again, as far as technical considerations allow, the coherent approach should consider large models where the robustness question looms less large. Notice that the method is quite specific about how the data should be analysed with any model, large or small; namely by calculation of the final probability using only the rules of the probability calculus. Analysis of the normal, linear model within either the standard or the Bayesian paradigm is straightforward. To extend the model and replace normality with t-distributions would cause substantial difficulties within the standard approach, if only because of the non-existence of sufficient statistics of small dimensionality, whereas anyone with adequate knowledge of the probability calculus and access to computing facilities could perform the arithmetic for any data set.

One difficulty with the above analysis of the adequacy of a model M is that it forces consideration of alternatives $\bar{M}$ to that model. This is not necessary in the approach using $E(x_2)$, except in so far as $E(\cdot)$ might informally be suggested by alternatives. Differently expressed, the coherent paradigm, by requiring a probability distribution over possible models, is essentially a method of *comparison* of models, not of the adequacy of a single model. (The point applies equally to values of random quantities.) The following example is designed to show why this difficulty is a real feature of any study of uncertainty, so that its avoidance using $E(\cdot)$, for instance, may be unsound.

A scientist has an observation which is a finite sequence of zeros and ones. His model for this is a Bernoulli sequence with chance θ. Contrast two cases. In the first he notices that the sequence has a 0 in every even place and 1 in every odd, an unlikely occurrence under his model. He seeks for an alternative model and, although he cannot specify it tightly, he sees one that could explain the data and has reasonable probability; both $p(X_1|\bar{M}, H)$ and $p(\bar{M}|H)$ for data X_1 and alternative $\bar{M}$ are not near zero. In the second case he notices that sequence has a 0 in every place whose order is composite and 1 against every prime. In this case no $\bar{M}$ satisfying the condition in the first case is available. Hence we have two pieces of data, X_1 and X_2, with $p(X_1|M, H) = p(X_2|M, H)$ very small, yet our reactions to them will be different because in one there exists an alternative with reasonable probability, which might be investigated, whereas in the other no such alternative exists and we can only conclude that an unusual observation x_2 has been observed. Actually all observations have something unusual about them: the key question is whether there is a reasonable alternative to explain the observation.

7. Cromwell's rule

Suppose that a random quantity X can only assume one of n values $x_1, x_2, \ldots, x_n$. An event is an example of this, where n is 2. If another random quantity Y is observed, the uncertainty about X will be updated by the usual Bayes' result

$$p(X|Y, H) \propto p(Y|X, H)p(X|H).$$

It immediately follows that if $p(x_i|H)$ is zero for any i then $p(x_i|Y, H)$ is also zero. Since Y is arbitrary, it follows that if $p(x_i|H) = 0$, no evidence whatsoever will alter the probability to any value other than zero. Whilst there is nothing in the Bayesian paradigm, as usually discussed, to rule out this possibility, it seems unreasonably rigid to fly in the face of all evidence, even when it supports strongly the possibility that $X = x_i$. We therefore suggest adding an additional requirement, namely that for quantities taking only a finite set of values, no probability be zero. With this addition it is easy to see that if indeed $X = x_i$, then data can eventually accumulate to make this almost certain. (The requirement might be called Cromwell's rule, since he suggested its equivalent when he advised the Church of Scotland to remember that it might be wrong.) In the extension to continuous X, where we are dealing with densities rather than probabilities, the equivalent requirement seems to be that the density nowhere vanishes.

Cromwell's rule is relevant when we consider the relationship between a Bayesian view of the world and the reality of that world that he learns by experience. We have seen that with a complete probabilistic description $p(X|H)$, with $X = (X_1, X_2)$ and X_2 observed, the experience is translated into $p(X_1|X_2, H)$, so that he need only update his probabilities according to the rules and no rethinking, only calculation, is needed. This may be unsatisfactory as the following example shows.

Let X consist of a long, perhaps infinite, sequence of ones and zeros and let $p(X|H)$ consist in saying that the sequence is a Bernoulli sequence with chance $\frac{1}{2}$ of one at any place. This is a coherent allocation of probabilities to X, assigning probability $(\frac{1}{2})^n$ to any subsequence of length n. Consequently, even if a sequence of 100 zeros is observed the probability of a 1 in the 101st place is still $\frac{1}{2}$. This scarcely seems reasonable and a second, coherent scientist watching this behaviour will discuss that view as "unscientific" in that it does not incorporate sensible reaction to the data.

The difficulty can be mitigated by judging the sequence to be, conditional on θ, Bernoulli with chance θ and ascribing to θ a density $p(\theta|H)$ which nowhere vanishes — unlike the earlier case where $p(\theta|H) = 0$ for all $\theta \neq \frac{1}{2}$. This is an example of a model as defined above. Now with any sequence observed, the probability of a 1 in the next place will be around r/n, where r is the number of ones in the sequence of length n. The second scientist might regard this appreciation of data as "scientific". But now suppose the observed sequence showed zeros and ones alternating, the probability for a 1 in the next place will be about $\frac{1}{2}$, since $r = \frac{1}{2}n$, whereas a more appealing value might be around 1 if the observed sequence ended in a 0, and 0 if ending in a 1. Hence even this Bayesian reacts unreasonably to some data.

This can be overcome by supposing the sequence to be a first-order Markov chain with parameters (θ, α), θ being as before. Now reaction to the alternating sequence will be reasonable but a sequence exhibiting only triplets 011, say, will be handled unreasonably. This can be countered by a second-order chain: and so on. Each case is defective when viewed in the light of the next case, in that it assigns zero densities. For example, the Bernoulli assignment gave zero density to all but one value of α in the Markov chain description. Cromwell's rule would have avoided the difficulties.

A realistic position seems to be that a coherent view must not assign density zero to any possibility (or alternatively zero probability to any open set); that at any stage it may work with a simplified model, such as a Markov chain, assigning some zero densities, having probability near one; but must be prepared to abandon it in the light of unexpected data that suggests the model may have lower probability than thought earlier. (This agrees with our discussion of models above.) This recipe seems to lead to reasonable appreciation of data. Cromwell's rule might therefore be added to our axiom system.

The rule is also related to the phenomenon of *calibration* that has been usefully studied by experimental psychologists: Lichtenstein *et al.* (1977) provide an excellent survey. A person is said to be calibrated if, after assigning probabilities to a long sequence of independent events, the frequency of events subsequently found to be true amongst all those assigned probability p, is p; and this for all p. Dawid (1982) has shown that *every* Bayesian is calibrated with probability one. For example, the scientist with a Bernoulli sequence of chance $\frac{1}{2}$ who does not learn from a long string of zeros is nevertheless calibrated. The "catch" in Dawid's result is that the probability referred to is the Bayesian's own probability, so that the adherent of chance $\frac{1}{2}$ has very low probability for the string of zeros. It is not true that one Bayesian will be calibrated with probability one according to Bayesian two. A desirable state of affairs would be that every Bayesian would attach probability one to another being calibrated. A requirement for this again seems to be Cromwell's rule so that if one gives probability greater than zero to some possibility, so will the other. Two such Bayesians will ultimately come to agree in the light of suitable data.

Notice that logic, as distinct from experience, can make a probability zero, or one. Thus lengthy, accurate calculations can establish the true value of the millionth digit of π, and Cromwell's rule does not apply.

8. Subjectivity and appraisal

Two related objections to the Bayesian paradigm are often raised. The first is that the procedure is subjective, the second that the probabilities are unknown. These have been much discussed but their frequent repetition suggests some more consideration might be desirable even though there is little new to say.

Consider two scientists contemplating a random quantity X. Then their probabilities for X might reasonably differ. The usual explanation for part, at least, of the difference is that their current states of information differ: that one is evaluating $p(X|H_1)$ and the other $p(X|H_2)$. Probability is a function of two arguments. It is, therefore, an important part of the Bayesian paradigm that information should be shared and that both scientists should contemplate X given $H = (H_1, H_2)$. We saw in Section 3 how this might be done coherently for both of them. It is an advantage of the paradigm that it provides the machinery for doing this sharing of knowledge. Even with the same information the two scientists might still have different probabilities $p_i(X|H)$, $i = 1, 2$. It is possible to argue that no two scientists have, or could have, identical Hs and that differences could still be ascribed to difference in parts of H not previously considered. An extension of this view is to argue that, on proper specification of X and H, all people would agree on $p(X|H)$, which may be thought of as a rational, or logical, view, and extends ordinary logic. Attempts to calculate $p(X|H)$, at least for simple situations, have not been totally successful but neither have they been total failures. Jeffreys (1967) is a protagonist for this view. Box and Tiao (1973) and Zellner (1971) both use the idea in their books. My own view is that the purely subjective view makes more sense. One disadvantage of the logical view is that it discourages serious contemplation of the probabilities—and even of X and H beyond their symbolic meanings—and adopts, perhaps uncritically, a logical probability. Most attempts at describing a logical probability reduce to considering the case where H is, if not empty, at least small, and $p(X|H)$ is a probability for X in a state of "ignorance". However, we are never ignorant: as long as the words mean something to us, we know something. At best an "ignorance" probability could only serve as a reference point for other, more realistic, ones.

On the subjective view there is no reason why $p_i(X|H)$ should agree for two scientists. However, it must be remembered that we are supposing the two scientists are coherent; that is, they have assessed other probabilities that combine together according to the rules of the probability calculus (Section 2). Much scientific opinion—and especially that based on orthodox statistical principles that deny the likelihood principle—is currently incoherent, and my conjecture is that some disagreement in views results from the scientist's failure to treat his data coherently. Certainly this is true of some significance test arguments and data-analytic techniques that do not assess evidence by consideration of alternative hypotheses (Section 6).

The message of this section is that two major impediments to agreement can be removed: use of different Hs and the failure of coherence. Yet still two scientists might disagree. But now remember that they have expressed their views in an easily understood way, namely by means of numbers (probabilities), and communication between them is easy because they speak a common tongue. Because of this it is possible for them to see where their principle differences lie and to concentrate attention on these. To do so may be enough to resolve differences. If it is not, then a possibility is to devise an experiment to reduce disagreement. For example, if one scientist attaches high probability to X_1 and another to X_2, the observation of Y, where $p(Y|X_1)$ and $p(Y|X_2)$ are agreed to differ substantially, essentially resolves the issue. It is necessary to invoke Cromwell's rule, for which one scientist denying with certainty what another scientist credits with being a possibility, is to deny the possibility of critical experimentation.

We see that the subjective approach has the great strength of encouraging cooperation and discussion amongst scientists and of suggesting new experiments. Always the argument comes back to coherence: to the fitting of judgements of uncertainty together. A scientist doing an experiment has been taught by statisticians that there is a unique, proper analysis of the data. When he attempts a Bayesian analysis he sees that this is not so: the analysis will depend on the "prior", on the probability before seeing the data. This will have come from previous evidence, perhaps of earlier experiments that the scientist did. The apparent merit of a unique analysis denies the proper appreciation of the relationship between experiments and the coherence of judgements concerning them. It denies part of learning from experience.

It is often maintained that the likelihood is objective even if the "prior" is not. This is not so: the probability from which it is derived is subjective, like any other probability assessment. There may be more agreement, usually because of the model feature (Section 4) of independence from H, given θ, but also because of exchangeability being agreed. But were coherence to be used more rigorously, we would have more information about reasonable likelihoods and have real evidence for normality or other assumptions.

The second objection is that the probabilities are unknown: loosely, what is the "prior"? The Bayesian paradigm purports to describe how a scientist would *wish* to behave, rather than how he *does* behave. To achieve this wish he needs to think probabilistically. Once convinced of this, he needs to assess probabilities, to supply actual numbers. Since he has not been doing this he needs to develop expertise and to develop tools to assist him. The objector is correct in asking his question: he is incorrect if he thinks that its answer is to come as revelation, without substantial studies by the scientist. He must think about the quantity, about ways of assessing it, and must face up to a research project. Scientists who encounter a measurement problem that they think is worth solution, do not say "I cannot do this", rather they launch a research programme to discover how to do it. So we should approach the assessment of probabilities.

Some work on the determination of probabilities has been done using scoring-rules and calibration—two topics mentioned above. Useful as both of these devices are, they fail to exploit the basic concept of coherence: they fail to see how one probability impinges on another. Ways of exploiting coherence have been suggested by Lindley et al. (1979). Essentially, the idea is to ask for *sets* of probabilities. Thus we might try $p(A|B)$, $p(A|\bar{B})$ and $p(B)$ for events A and B. The coherent assessor is then committed to $p(B|A)$, indeed to all uncertainties concerning A and B. He may feel that the calculated $p(B|A)$ is unsatisfactory. He can then modify any or all of the original triplet of values to attain coherence and to agree with his total appreciation of the uncertainties surrounding the events.

Neither of the objections dealt with in this section recognize the power of coherence: the ability of that tool to fit opinions concerning uncertainty together in an unobjectionable way. No other procedure besides the Bayesian method can do this: for that method follows from reasonable assumptions of behaviour.

References

Box, G. E. P. (1980). Sampling and Bayes' inference in scientific modeling and robustness, *Journal of the Royal Statistical Society* A, **143**, 383–430.

Box, G. E. P. and Tiao, G. C. (1973). *Bayesian Inference in Statistical Analysis*. Addison-Wesley, Reading, Massachusetts.

Dawid, A. P. (1979). Conditional independence in statistical theory, *Journal of the Royal Statistical Society* B, **41**, 1–31.

Dawid, A. P. (1982). The well-calibrated Bayesian, *Journal of the American Statistical Association* (to appear).

De Finetti, B. (1972). *Probability, Induction and Statistics: The Art of Guessing*. Wiley, New York.

De Finetti, B. (1974/75). *Theory of Probability*, 2 Volumes. Wiley, New York.

DeGroot, M. H. (1970). *Optimal Statistical Decisions*. McGraw-Hill, New York.

Jeffreys, H. (1967). *Theory of Probability*. Clarendon Press, Oxford.

Lichtenstein, S., Fischhoff, B. and Phillips, L. D. (1977). Calibration of probabilities: The state of the art. In *Decision-making and Change in Human Affairs* (H. Jungermann and G. de Zeeuw, eds), 275–324. Reidel, Dordrecht.

Lindley, D. V. (1971). *Bayesian Statistics: A Review*. SIAM, Philadelphia.
Lindley, D. V. (1978). The Bayesian approach, *Scandinavian Journal of Statistics*, **5**, 1–26.
Lindley, D. V. and Novick, M. R. (1981). The role of exchangeability in inference, *Annals of Statistics*, **9**, 45–58.
Lindley, D. V., Tversky, A. and Brown, R. V. (1979). On the reconciliation of probability assessments, *Journal of the Royal Statistical Society* A, **142**, 146–180.
Ramsey, F. P. (1964). Truth and probability. In *Studies in Subjective Probability* (H. E. Kyburg, Jr. and H. E. Smokler, eds), 61–92. Wiley, New York.
Savage, L. J. (1954). *The Foundations of Statistics*. Wiley, New York.
Young, A. S. (1977). A Bayesian approach to prediction using polynomials, *Biometrika*, **64**, 309–317.
Zellner, A. (1971). *An Introduction to Bayesian Inference in Econometrics*. Wiley, New York.

5

CONCOMITANTS OF ORDER STATISTICS: THEORY AND APPLICATIONS

H. A. David

Abstract

Let (X_i, Y_i) $(i = 1, 2, \ldots, n)$ be independent pairs of variates having an absolutely continuous joint distribution. If $X_{r:n}$ denotes the rth ordered X-variate, then the Y-variate paired with $X_{r:n}$ is termed the concomitant of the rth order statistic and denoted by $Y_{[r:n]}$.

Exact and asymptotic distribution theory for $Y_{[r:n]}$ is developed when X_i and Y_i are linked by a linear regression model. The results are applied to (a) procedures based on the selection of individuals with the largest X-values, (b) estimation of regression and correlation coefficients, (c) estimation for censored bivariate data and (d) double sampling when the Xs represent inexpensive auxiliary measurements which can be used to choose a small number of individuals for whom expensive Y-measurements are to be made for the estimation of EY.

In the remaining parts of the paper the linear regression model is dispensed with. General distributional results are obtained for the $Y_{[r:n]}$ and also for the rank of $Y_{[r:n]}$ among the Ys. Further applications are given.

---

1. Introduction and basic distribution theory

Let (X_i, Y_i) $(i = 1, 2, \ldots, n)$ be n independent pairs of variates from an absolutely continuous bivariate distribution with CDF $P(x, y)$. If the pairs are

Research supported by the US Army Research Office.

ordered by their X-variates, then the Y-variate associated with $X_{r:n}$ will be denoted by $Y_{[r:n]}$ and termed the *concomitant of the rth order statistic* (David, 1973). These concomitants are of interest in selection problems and also in a variety of estimation procedures. The purpose of this paper is to provide an integrated account of the subject.

Suppose that X_i and Y_i ($i = 1, 2, \ldots, n$) have means μ_X, μ_Y, variances σ_X^2, σ_Y^2, and are linked by the linear regression model ($|\rho| < 1$)

$$Y_i = \mu_Y + \rho \frac{\sigma_Y}{\sigma_X}(X_i - \mu_X) + Z_i, \tag{1.1}$$

where the X_i and the Z_i are mutually independent. Then from (1.1) it follows that $EZ_i = 0$, var $Z_i = \sigma_Y^2(1-\rho^2)$ and $\rho = \text{corr}(X, Y)$. In the special case when the X_i and Z_i are normal, X_i and Y_i are bivariate normal. Ordering on the Xs we have for $r = 1, 2, \ldots, n$

$$Y_{[r:n]} = \mu_Y + \rho \frac{\sigma_Y}{\sigma_X}(X_{r:n} - \mu_X) + Z_{[r]}, \tag{1.2}$$

where $Z_{[r]}$ denotes the particular Z_i associated with $X_{r:n}$. In view of the independence of the X_i and the Z_i we see that the set of $X_{r:n}$ is independent of the $Z_{[r]}$, the latter being mutually independent each with the same distribution as Z_i. Setting

$$\alpha_{r:n} = E\left(\frac{X_{r:n} - \mu_X}{\sigma_X}\right) \quad \text{and} \quad \beta_{rs:n} = \text{cov}\left(\frac{X_{r:n} - \mu_X}{\sigma_X}, \frac{X_{s:n} - \mu_X}{\sigma_X}\right),$$

$r, s = 1, 2, \ldots, n$, we have from (1.2)

$$\begin{aligned} EY_{[r:n]} &= \mu_Y + \rho \sigma_Y \alpha_{r:n}, \\ \text{var } Y_{[r:n]} &= \sigma_Y^2 (\rho^2 \beta_{rr:n} + 1 - \rho^2), \\ \text{cov}(X_{r:n}, Y_{[s:n]}) &= \rho \sigma_X \sigma_Y \beta_{rs:n}, \\ \text{cov}(Y_{[r:n]}, Y_{[s:n]}) &= \rho^2 \sigma_Y^2 \beta_{rs:n}, \quad r \neq s. \end{aligned} \tag{1.3}$$

In the bivariate normal case eqns (1.3) were given by Watterson (1959).

The distribution of $Y_{[r:n]}$ for finite n follows directly from (1.2). Now suppose that $n \to \infty$ with $r/n \to \lambda$, a constant. We must distinguish between the quantile case ($0 < \lambda < 1$) and the extreme-value case ($\lambda = 0$ or 1). For ease of writing take $\mu_X = \mu_Y = 0$, $\sigma_X = \sigma_Y = 1$ in (1.2) so that

$$Y_{[r:n]} = \rho X_{r:n} + Z_{[r]}. \tag{1.4}$$

Since in the quantile case $X_{r:n}$ converges in probability to $P^{-1}(\lambda)$ (where P is written for P_X), we see at once that the asymptotic distribution of $Y_{[r:n]} - \rho P^{-1}(\lambda)$ coincides with the distribution of Z.

The situation is more complicated in the extreme-value case. For definiteness take $r = n$, but the results of this paragraph will hold for all upper extremes, i.e. $r = n - m + 1$, with m constant (and corresponding results hold for lower extremes). There is no difficulty when the distribution of X is limited on the right, say at x_0, since then $X_{n:n}$ converges in probability to x_0. Suppose therefore that $P(x) < 1$ for every finite x. Then there exists (Gnedenko, 1943) a sequence of constants $\{A_n\}$ such that $\{X_{n:n} - A_n\}$ converges to 0 in probability iff

$$\lim_{x \to \infty} \frac{1 - P(x + \varepsilon)}{1 - P(x)} = 0 \tag{1.5}$$

for every $\varepsilon > 0$. These conditions are satisfied, in particular, when X and Y are bivariate normal, with $A_n = (2 \log n)^{1/2}$. From (1.4) we have, therefore, that $Y_{[r:n]} - EY_{[r:n]}$ is asymptotically normal $N(0, 1 - \rho^2)$, where

$$EY_{[r:n]} \sim \rho P^{-1}(\lambda) \qquad 0 < \lambda < 1,$$
$$\sim \rho (2 \log n)^{1/2} \qquad \lambda = 1,$$
$$\sim -\rho (2 \log n)^{1/2} \qquad \lambda = 0.$$

Let us look at upper extremes more generally. If $P(x)$ is such that $(X_{r:n} - b_n)/a_n$ has limiting CDF $\Lambda(x)$ for suitable choices of a_n and b_n, then the stochastic growth of $X_{r:n}$ with n is determined primarily by the growth of a_n. It is well known that there are three possible limiting distributions of $X_{r:n}$ for $r = n - m + 1$, viz.

$$\Lambda_i^{(m)}(x) = \Lambda_i(x) \sum_{j=0}^{m-1} \lambda_i^j(x)/j!,$$

where

$$\lambda_i(x) = -\log \Lambda_i(x), \qquad i = 1, 2, 3$$

with

$$\Lambda_1(x) = 0, \qquad\qquad x \leqslant 0, \alpha > 0$$
$$= \exp(-x^{-\alpha}), \qquad x > 0;$$
$$\Lambda_2(x) = \exp[-(-x)^{-\alpha}], \qquad x \leqslant 0, \alpha > 0,$$
$$= 1, \qquad\qquad x > 0;$$
$$\Lambda_3(x) = \exp(-e^{-x}), \qquad -\infty < x < \infty.$$

For $P(x)$ in the domain of attraction of $\Lambda_1(x)$ (Type I) one has

$$\lim_{n \to \infty} \Pr\{X_{r:n} \leqslant a_n x\} = \Lambda_1^{(m)}(x),$$

where a_n may be taken as

$$a_n = P^{-1}(1-1/n).$$

Thus $a_n \to \infty$ as $n \to \infty$ and $Y_{[r:n]}$ will from (1.4) behave asymptotically as $\rho X_{r:n}$. In the Type II case we have already noted that $Y_{[r:n]}$ behaves asymptotically as $\rho x_0 + Z$, where here $x_0 = 0$. For the Type III case, $\lim a_n$ may assume any non-negative value. If $\lim a_n$ is positive and finite, then the two terms on the right of (1.4) are of the same order in variance; more specifically, the convolution of the distribution of Z and the asymptotic distribution of $\rho(X_{r:n} - b_n)$ gives the asymptotic distribution of $Y_{[r:n]} - \rho b_n$. Of course, if $\lim a_n = 0$ in this Type III situation, which is the case for the bivariate normal, then $Y_{[r:n]} - \rho b_n$ behaves asymptotically as Z.

2. Applications in selection and estimation procedures

Selection

Consider the following selection procedure. From n individuals the $k(<n)$ with the largest X-values are to be chosen. What can we predict about their performance as measured by an associated characteristic Y? Clearly, we are really asking about the behavior of $Y_{[r:n]}$ for $r = n-k+1,\ldots,n$. This basic application of concomitants of order statistics long preceded the coining of the term.

Often one is interested in the *selection differential* of the $Y_{[r:n]}$, viz.

$$D_{[k:n]} = \frac{1}{k} \sum_{i=n-k+1}^{n} \frac{Y_{[i:n]} - \mu_Y}{\sigma_Y}.$$

If $D_{k:n}$ is the selection differential of the $X_{r:n}$, i.e.

$$D_{k:n} = \frac{1}{k} \sum_{i=n-k+1}^{n} \frac{X_{i:n} - \mu_X}{\sigma_X},$$

then, for the linear regression model, we have from (1.2)

$$D_{[k:n]} = \rho D_{k:n} + \frac{1}{k} \sum_{i=n-k+1}^{n} \frac{Z_{[i]}}{\sigma_Y},$$

and hence

$$ED_{[k:n]} = \rho ED_{k:n},$$

$$\operatorname{var} D_{[k:n]} = \rho^2 \operatorname{var} D_{k:n} + \frac{1}{k}(1-\rho^2). \tag{2.1}$$

Thus the gain due to selection on X is attenuated for Y by the factor $|\rho|$. Exact values of the mean and variance of $D_{[k:n]}$ can from (2.1) be obtained with the help of tables of the first two moments of the order statistics. In the bivariate normal case approximations have been developed by Burrows (1972, 1975).

Estimation of regression and correlation coefficient

If the regression of Y on the non-stochastic variable x is linear, viz.

$$E(Y|x) = \alpha + \beta x, \qquad (2.2)$$

then β may be estimated by the ratio statistic

$$b' = \frac{\bar{Y}'_{[k:n]} - \bar{Y}_{[k:n]}}{\bar{x}'_{k:n} - \bar{x}_{k:n}}, \qquad (2.3)$$

where

$$\bar{x}'_{k:n} = \frac{1}{k}\sum_{i=1}^{k} x_{n+1-i:n}, \qquad \bar{x}_{k:n} = \frac{1}{k}\sum_{i=1}^{k} x_{i:n},$$

and

$$\bar{Y}'_{[k:n]} = \frac{1}{k}\sum_{i=1}^{k} Y_{[n+1-i:n]}, \qquad \bar{Y}_{[k:n]} = \frac{1}{k}\sum_{i=1}^{k} Y_{[i:n]}.$$

If X is stochastic, we may interpret (2.2) as conditional on $X = x$ and have from (2.3)

$$E(b'|x_1, x_2, \ldots, x_n) = \beta. \qquad (2.4)$$

Since (2.4) holds whatever the x_i, it also holds unconditionally; that is,

$$B' = \frac{\bar{Y}'_{[k:n]} - \bar{Y}_{[k:n]}}{\bar{X}'_{k:n} - \bar{X}_{k:n}} \qquad (2.5)$$

is also an unbiased estimator of β. Note that this result does not require either the Xs or the Ys to be identically distributed or even to be independent. Barton and Casley (1958) show that B' has an efficiency of 75–80% when $(X_i, Y_i), i = 1, 2, \ldots, n$, is a random sample from a bivariate normal, provided k is chosen as about $0.27n$.

It is possible, but less satisfactory, to estimate ρ in a similar manner. See Tsukibayashi (1962) and Barnett et al. (1976).

Estimation for censored bivariate data

Concomitants of order statistics arise very naturally when bivariate data $(x_i, y_i), i = 1, 2, \ldots, n$, are censored. Three kinds of censoring may usefully be distinguished (Watterson, 1959): (a) censoring of certain $x_{i:n}$ and of the corresponding $y_{[i:n]}$; (b) censoring of certain $y_{[i:n]}$ only; and (c) censoring of certain $x_{i:n}$ only. For example, (b) occurs when the $x_{i:n}$ $(i=1,2,\ldots,n)$ are entrance scores and the $y_{[i:n]}$ $(i=k+1, k+2,\ldots,n)$ later scores of the successful candidates. On the other hand, (c) applies in a life test terminated after $n-k$ failures when measurements on some associated variable are available for all n items.

In the bivariate normal case Watterson (1959) obtains various linear estimates which are unbiased but have variances depending on ρ. For case (a) maximum likelihood methods are used by Harrell and Sen (1979).

Double sampling

Suppose we wish to estimate μ_Y when Y is expensive to measure. If an inexpensive auxiliary variable X is available, then taking n measurements on X, we can use their ordering to make $k(<n)$ expensive measurements $Y_{[r_j:n]}$ $(j=1,2,\ldots,k)$. A simple estimator of μ_Y is their average $\bar{Y}_{[r:n]}$, say, which under the model (1.1) is evidently unbiased for any symmetric distribution of X if

$$r_{k+1-j} = n+1-r_j \quad j=1,2,\ldots,\tfrac{1}{2}k \ (k \text{ even}), j=1,2,\ldots,\tfrac{1}{2}(k+1) \ (k \text{ odd}).$$

Moreover, since from (1.2),

$$\bar{Y}_{[r:n]} = \mu_Y + \rho \frac{\sigma_Y}{\sigma_X}(\bar{X}_{r:n}-\mu_X)+\bar{Z},$$

where $\bar{X}_{r:n}$ is the average of the k $X_{r_j:n}$ and $\bar{Z}$ of the k $Z_{[r_j]}$, we have

$$\operatorname{var}\left(\frac{\bar{Y}_{[r:n]}}{\sigma_Y}\right) = \rho^2 \operatorname{var}\left(\frac{\bar{X}_{r:n}}{\sigma_X}\right)+\frac{1}{k}(1-\rho^2). \tag{2.6}$$

Thus the ranks r_j minimizing var $\bar{X}_{r:n}$ also minimize var $\bar{Y}_{[r:n]}$, whatever the value of ρ. The simple form of optimal spacing involved in minimizing var $\bar{X}_{r:n}$ was studied by Mosteller (1946). The (asymptotically) optimal ranks are given as the integral parts of $n\lambda_j+1$, where $0<\lambda_1<\lambda_2<\cdots<\lambda_k<1$, and the λ_j have been tabulated in the normal case for $k \leq 10$; roughly $\lambda_j = (j-\tfrac{1}{2})/k$. In practice, the first term on the right of (2.6), which vanishes as $n\to\infty$, is small compared with the second; asymptotically $\bar{Y}_{r:n}$ has a variance $(1-\rho^2)$ times that of the mean of k randomly chosen Ys.

The foregoing and related procedures are treated by O'Connell and David (1976). An interesting method with a similar aim and also using concomitants is *ranked-set sampling*. This was introduced by McIntyre (1952) and further investigated by Dell and Clutter (1972). Here the sample size n is taken to be $n=k^2$ (or a multiple of k^2) and k subsamples each of size k are formed. All X-measurements are made (e.g. visual rankings) but only one Y is measured per subsample, namely $Y_{[j:k]}$ in the jth subsample $(j=1,2,\ldots,k)$. McIntyre's estimator of μ_Y is then the average of the $Y_{[j:k]}$. This is easily seen to be unbiased for any parent distribution, but is less efficient than $\bar{Y}_{[r:n]}$ in the normal case. See also David (1981). Ranked-set sampling has recently been applied to the estimation of variance by Stokes (1980).

3. The rank of $Y_{[r:n]}$

Suppose an individual A_i ($i = 1, 2, \ldots, n$) ranks rth on the x-measurement (e.g. first test). In this section we study the following two questions: (a) What is the probability that A_i will rank sth on the y-measurement (e.g. second test)? (b) What is A_is expected rank on the y-measurement?

Let $R_{r,n}$ denote the rank of $Y_{[r:n]}$ among the n Ys, i.e.

$$R_{r,n} = \sum_{i=1}^{n} I(Y_{[r:n]} - Y_i), \qquad (3.1)$$

where

$$I(x) = 1 \quad \text{if } x \geq 0,$$
$$= 0 \quad \text{if } x < 0.$$

For (a) we require $\pi_{rs} = \Pr\{R_{r,n} = s\}$. Consider first π_{nn}. We have

$$\pi_{nn} = \sum_{i=1}^{n} \Pr\{X_i = X_{n:n}, Y_i = Y_{n:n}\}$$
$$= n \Pr\{X_n = X_{n:n}, Y_n = Y_{n:n}\}$$
$$= n \Pr\{X_1 < X_n, \ldots, X_{n-1} < X_n; Y_1 < Y_n, \ldots, Y_{n-1} < Y_n\}$$
$$= n \int_{-\infty}^{\infty} \int_{-\infty}^{\infty} [\Pr\{X < x, Y < y\}]^{n-1} \, dP(x, y)$$

upon conditionalizing on X_n, Y_n. It is not difficult to extend this argument to finding π_{rs}. Table 1 is an extract from a table in David et al. (1977) for the case when X and Y are bivariate normal.

Example. Suppose that the scores of candidates taking two tests are bivariate normal with $\rho = 0.8$. Out of 9 candidates taking the first (screening) test the top k are selected and given the second test. What is the smallest value of k ensuring with probability at least 0.9 that the best of the 9 candidates, as judged by the second test, is included among the k selected?

We require the smallest k such that

$$\pi_{99} + \pi_{89} + \cdots + \pi_{10-k,9} \geq 0.9$$

Table 1. $\pi_{rs} = \Pr\{R_{r,n} = s\}$ as a function of ρ for $n = 9$

r	s	ρ									
		0.10	0.20	0.30	0.40	0.50	0.60	0.70	0.80	0.90	0.95
9	9	0.1407	0.1746	0.2133	0.2576	0.3087	0.3686	0.4404	0.5306	0.6564	0.7510
	8	0.1285	0.1459	0.1631	0.1797	0.1952	0.2087	0.2185	0.2207	0.2033	0.1725
	7	0.1211	0.1296	0.1363	0.1408	0.1424	0.1401	0.1321	0.1152	0.0817	0.0523
	6	0.1152	0.1173	0.1171	0.1143	0.1085	0.0989	0.0846	0.0640	0.0350	0.0169
	5	0.1100	0.1069	0.1015	0.0938	0.0836	0.0706	0.0546	0.0357	0.0149	0.0053
	4	0.1051	0.0973	0.0877	0.0765	0.0638	0.0497	0.0345	0.0192	0.0059	0.0015
	3	0.0999	0.0877	0.0747	0.0611	0.0472	0.0334	0.0205	0.0095	0.0021	0.0004
	2	0.0939	0.0773	0.0612	0.0462	0.0324	0.0204	0.0107	0.0040	0.0006	0.0001
	1	0.0856	0.0635	0.0451	0.0300	0.0183	0.0097	0.0041	0.0011	0.0001	0.0000

i.e.
$$\pi_{99} + \pi_{98} + \cdots + \pi_{9,10-k} \geq 0.9.$$

Since, from the column for $\rho = 0.8$, we have
$$0.5306 + 0.2207 + 0.1152 + 0.0640 = 0.9305$$
the required value is $k = 4$.

The expected value of $R_{r,n}$ may be obtained directly by the following characteristic order statistics argument. Let $X_{r:n-1}$ and $X_{r:n}$ denote the rth order statistic for $X_1, X_2, \ldots, X_{n-1}$ and $X_1, X_2, \ldots, X_{n-1}, X_n$, respectively. Then clearly

$$\begin{array}{cc} X_{r-1:n-1} & X_{r:n-1} \\ \vdash & \vdash \end{array}$$

$$\begin{aligned} X_{r:n} &= X_{r-1:n-1} & \text{if } X_n < X_{r-1:n-1} \\ &= X_{r:n-1} & \text{if } X_n > X_{r:n-1} \\ &= X_n & \text{if } X_{r-1:n-1} < X_n < X_{r:n-1} \end{aligned} \qquad (3.2)$$

From (3.1) we have
$$E(R_{r,n}) = n \Pr\{Y_{[r:n]} \geq Y_n\}, \qquad (3.3)$$
where corresponding to (3.2)
$$\Pr\{Y_{[r:n]} \geq Y_n\} = \Pr\{X_n < X_{r-1:n-1}, Y_n < Y_{[r-1:n-1]}\}$$
$$+ \Pr\{X_n > X_{r:n-1}, Y_n < Y_{[r:n-1]}\} + \frac{1}{n}.$$

On noting that the joint PDF $f_{X_{r-1:n-1}, Y_{[r-1:n-1]}}(x, y)$ may be written as $f_{r-1:n-1}(x) \cdot f(y|x)$ we now obtain

$$E(R_{r,n}) = 1 + n \int_{-\infty}^{\infty} \int_{-\infty}^{\infty} f(y|x) [\Pr\{X < x, Y < y\} f_{r-1:n-1}(x)$$
$$+ \Pr\{X > x, Y < y\} f_{r:n-1}(x)] \, dy \, dx.$$

In the bivariate normal case some further simplification is possible. It turns out (David et al., 1977) that, for $r/(n+1) \to \lambda$ $(0 < \lambda < 1)$ as $n \to \infty$, $E[R_{r,n}/(n+1)]$ is quite well approximated, even for n as small as 9, by its asymptotic value

$$\lim_{n \to \infty} E\left(\frac{R_{r,n}}{n+1}\right) = \Phi\left(\frac{\rho \Phi^{-1}(\lambda)}{(2-\rho^2)^{1/2}}\right),$$

where Φ denotes the standard normal CDF. It can also be shown (David and Galambos, 1974; David et al., 1977) that for $0 \leq u \leq 1$

$$\lim_{n\to\infty} \Pr\{R_{r,n} \leq nu\} = \Phi\left(\frac{\Phi^{-1}(u) - \rho\Phi^{-1}(\lambda)}{(1-\rho^2)^{1/2}}\right).$$

For the general bivariate case Yang (1977) obtains correspondingly the intuitively appealing result

$$\lim_{n\to\infty} \Pr\{R_{r,n} \leq nu\} = \Pr\{Y \leq P_Y^{-1}(u) | X = P_X^{-1}(\lambda)\}$$

4. General distribution theory and further applications

It is possible to develop a general distribution theory of concomitants which assumes only that (X_i, Y_i) $(i = 1, 2, \ldots, n)$ are independently drawn from a common absolutely continuous bivariate population. The linear regression model (1.1) has allowed us to obtain a number of rather explicit results but was already dispensed with in the preceding section.

For $1 \leq r_1 < r_2 < \cdots < r_k \leq n$ it is clear that quite generally the $Y_{[r_j:n]}$ ($j = 1, 2, \ldots, k$) are conditionally independent given $X_{r_j:n} = x_j$ ($j = 1, 2, \ldots, k$). The joint conditional PDF may be written as $\prod_{j=1}^{k} f(y_j | x_j)$. It follows that

$$f_{Y_{[r_1:n]}, \ldots, Y_{[r_k:n]}}(y_1, \ldots, y_k) = \int_{-\infty}^{\infty} \int_{-\infty}^{x_k} \cdots \int_{-\infty}^{x_2} f_{X_{r_1:n}, \ldots, X_{r_k:n}}(x_1, \ldots, x_k) \prod_{j=1}^{k} [f(y_j | x_j) \, dx_j].$$

Put $m(x) = E(Y|X = x)$ and $\sigma^2(x) = \text{var}(Y|X = x)$ (Bhattacharya, 1974). It follows (Yang, 1977), in generalization of (1.3), that

$$E(Y_{[r:n]}) = E[m(X_{r:n})],$$

$$\text{var}(Y_{[r:n]}) = \text{var}[m(X_{r:n})] + E[\sigma^2(X_{r:n})],$$

$$\text{cov}(X_{r:n}, Y_{[s:n]}) = \text{cov}[X_{r:n}, m(X_{s:n})],$$

$$\text{cov}(Y_{[r:n]}, Y_{[s:n]}) = \text{cov}[m(X_{r:n}), m(X_{s:n})], \quad r \neq s.$$

Simple asymptotic results can be obtained when, as in the quantile case, the $X_{r:n}$ converge in probability to a finite value, say $P^{-1}(\lambda_i)$. Thus from

$$\Pr\{Y_{[r:n]} \leq y\} = \int \Pr\{Y_i \leq y | X_i = x\} \, dF_{X_{r:n}}(x)$$

we have from the Helly–Bray Theorem, since $\Pr\{Y_i \leq y | X_i = x\}$ is a bounded continuous function, that

$$\lim_{n\to\infty} \Pr\{Y_{[r:n]} \leq y\} = \Pr\{Y_i \leq y | X_i = P^{-1}(\lambda)\}.$$

In fact, this argument generalizes immediately to show that a finite number k of the $Y_{[r_j:n]}$ are asymptotically independent. More precisely (Yang, 1977), if

$r_j/n \to \lambda_j$ $(j = 1, 2, \ldots, k)$ as $n \to \infty$ with $0 < \lambda_j < 1$, then

$$\lim_{n \to \infty} \Pr\{Y_{[r_1:n]} \leqslant y_1, \ldots, Y_{[r_k:n]} \leqslant y_k\} = \prod_{j=1}^{k} \Pr\{Y_j \leqslant y_j | X_j = P^{-1}(\lambda_j)\}.$$

Bhattacharya (1974)‡ considers the limiting behavior of the sample paths of

$$\sum_{i=1}^{k} [Y_{[i:n]} - m(X_{i:n})] \tag{4.1}$$

for $k = 1, 2, \ldots, n$ and arrives at the following large sample test of a specified regression function: For a level α test of $m(x) - m_0(x)$ reject the null hypothesis if

$$W_n \geqslant \Phi^{-1}(1-\alpha) \quad \text{or} \quad W_n \leqslant \Phi^{-1}(\alpha),$$

where Φ denotes the standard normal CDF and W_n may be written as

$$\frac{\sum_{i=1}^{n-1} (n-i)[Y_{[i:n]} - m_0(X_{i:n})]}{\sum_{i=1}^{n-1} (n^2 - i^2)[Y_{[i:n]} - m_0(X_{i:n})]}.$$

Tests for the equality of two regression functions are given in Bhattacharya (1976). In the last paper Bhattacharya studies specifically the asymptotic behavior of what is essentially the selection differential $D_{[k:n]}$ of Section 2. A fuller treatment of this subject using several different approaches is given by Nagaraja (1980). The asymptotic distribution of general linear functions (with bounded smooth weight functions) of the $Y_{[i:n]}$ has been studied by Yang (1979).

References

Barnett, V., Green, P. J., and Robinson, A. (1976). Concomitants and correlation estimates, *Biometrika*, **63**, 323–328.

Barton, D. E. and Casley, D. J. (1958). A quick estimate of the regression coefficient, *Biometrika*, **45**, 431–435.

Bhattacharya, P. K. (1974). Convergence of sample paths of normalized sums of induced order statistics, *Annals of Statistics*, **2**, 1034–1039.

Bhattacharya, P. K. (1976). An invariance principle in regression analysis, *Annals of Statistics*, **4**, 621–624.

Burrows, P. M. (1972). Expected selection differentials for directional selection, *Biometrics*, **28**, 1091–1100.

‡This author, whose emphasis is different from ours, uses the term *induced order statistics* for concomitants of order statistics.

Burrows, P. M. (1975). Variances of selection differentials in normal samples, *Biometrics*, **31**, 125–133.
David, H. A. (1973). Concomitants of order statistics, *Bulletin of the International Statistical Institute*, **45**, 295–300.
David, H. A. (1981). *Order Statistics*, 2nd edn. Wiley, New York.
David, H. A. and Galambos, J. (1974). The asymptotic theory of concomitants of order statistics, *Journal of Applied Probability*, **11**, 762–770.
David, H. A., O'Connell, M. J. and Yang, S. S. (1977). Distribution and expected value of the rank of a concomitant of an order statistic, *Annals of Statistics*, **5**, 216–223.
Dell, T. R. and Clutter, J. L. (1972). Ranked set sampling theory with order statistics background, *Biometrics*, **28**, 545–555.
Gnedenko, B. (1943). Sur la distribution limite du terme maximum d'une série aléatoire, *Annals of Mathematics*, **44**, 423–453.
Harrell, F. E. and Sen, P. K. (1979). Statistical inference for censored bivariate normal distributions based on induced order statistics, *Biometrika*, **66**, 293–298.
McIntyre, G. A. (1952). A method of unbiased selective sampling, using ranked sets, *Australian Journal of Agricultural Research*, **3**, 385–390.
Mosteller, F. (1946). On some useful "inefficient" statistics, *Annals of Mathematical Statistics*, **17**, 377–408.
Nagaraja, H. N. (1980). Contributions to the Theory of the Selection Differential. Ph.D. thesis, Iowa State University.
O'Connell, M. J. and David, H. A. (1976). Order statistics and their concomitants in some double sampling situations. In *Essays in Probability and Statistics* (S. Ikeda et al., eds), 451–466. Shinko Tsusho, Tokyo.
Stokes, S. L. (1980). Estimation of variance using judgment ordered ranked set samples, *Biometrics*, **36**, 35–42.
Tsukibayashi, S. (1962). Estimation of bivariate parameters based on range, *Reports of Statistical Applied Research* JUSE, **9**, 10–23.
Watterson, G. A. (1959). Linear estimation in censored samples from multivariate normal populations, *Annals of Mathematical Statistics*, **30**, 814–824.
Yang, S. S. (1977). General distribution theory of the concomitants of order statistics, *Annals of Statistics*, **5**, 996–1002.
Yang, S. S. (1981). Linear functions of concomitants of order statistics with application to nonparametric estimation of a regression function, *Journal of the American Statistical Association*, **76**, 658–662.

6

DECISION AND MODELLING FOR EXTREMES

J. Tiago de Oliveira

Abstract

The paper contains, essentially, two parts.

In the first part, a method for selection between the three models for univariate independent extremes is exposed (Weibull, Gumbel and Fréchet). The solution of the trilemma is centred around Gumbel distribution (which has, in fact, a central position in Jenkinson's form). The method—whose asymptotic behaviour is studied—solves, first, the problem when location and dispersion parameters are known and, later, when they are unknown.

In the second part a new model for bivariate extremes—the natural model—is introduced, extending the already now classical biextremal model, and its dual. The basic properties are studied. Estimators for two parameters (which define the support of the random pair), other than the margin's location and dispersion parameters, are given.

1. Introduction

The purpose of this paper is two-directional. After some revision of what can be considered the classical results for statistical extremes, plus some recent results, we will deal with the choice of a statistical univariate extreme model and with the natural model for bivariate extremes. For details see Tiago de Oliveira (1975, 1980a, b) and references therein.

As is well known, the limiting distributions of maxima in the IID (independent and identically distributed) case, in reduced form (no location or dispersion parameters), are:

$$\Lambda(z) = \exp(-e^{-z}), \qquad -\infty < z < +\infty$$

(Gumbel distribution),

$$\Phi_\alpha(z) = 0 \quad \text{if } z \leq 0,$$
$$= \exp(-z^{-\alpha}) \quad \text{if } z \geq 0; \alpha > 0$$

(Fréchet distribution),

$$\Psi_\alpha(z) = \exp(-(-z)^\alpha) \quad \text{if } z \leq 0; \alpha > 0,$$
$$= 1 \quad \text{if } z \geq 0$$

(Weibull distribution).

It is well known that the limiting distributions of minima in the IID case are $1 - \Lambda(-z)$, $1 - \Phi_\alpha(-z)$ and $1 - \Psi_\alpha(-z)$, in reduced form. Recall that the limiting distributions of maxima and minima, for the same initial distribution, are not necessarily the corresponding ones, as the limiting behaviour depends on the tails (right tail for maxima, left tail for minima) and the two tails may behave differently.

Let us recall that probabilistic theory is enlarging the justification of the use of classical limiting results, whose details can be seen in Gnedenko (1943), Gumbel (1958) and Galambos (1978). The weakening of IID conditions can be seen in Juncosa (1949), with regard to the ID assumption, and in Watson (1954), Berman (1962, 1964), Newell (1964), Loynes (1965), Leadbetter (1974) and Turkman (1979), with respect to the independence condition. A recent paper by Tiago de Oliveira (1978) has a new approach where both conditions are weakened. In those papers, under different assumptions, the already given forms of limiting distributions are obtained.

This justifies the widespread use of Gumbel, Fréchet and Weibull distributions when dealing with samples of maxima and minima.

2. Some basic decision results for Gumbel distribution

In applications we will have to introduce location (λ) and dispersion ($\delta > 0$) parameters. Thus Gumbel distribution will have the form $\Lambda[(x-\lambda)/\delta]$. We will briefly sketch point estimation for this distribution which will be important in the sequel; for more details see Tiago de Oliveira (1975).

For an independent sample $(x_1, \ldots, x_n)$, the likelihood having the expression

$$\mathscr{L}(\lambda, \delta | x_i) = \frac{1}{\delta^n} \exp\left[-\frac{n(\bar{x}-\lambda)}{\delta}\right] \exp\left[-\Sigma \exp\left(-\frac{x_i-\lambda}{\delta}\right)\right],$$

the ML-estimators are, after some algebra,

$$\hat{\lambda} = -\hat{\delta} \log\{[\Sigma \exp(-x_i/\hat{\delta})]/n\},$$
$$\bar{x} = \hat{\delta} + \Sigma x_i \exp(-x_i/\hat{\delta})/\Sigma \exp(-x_i/\hat{\delta}).$$

The second equation can be solved by the Newton–Raphson method. Using the fact that the mean value of $\Lambda(z)$ is γ ($=0.57722\ldots$, the Euler constant) and the variance is $\pi^2/6$ the method of moments (generally the mean value and variance are $\lambda+\gamma\delta$ and $(\pi^2/6)\delta^2$) leads to the first estimate of δ given by $S^2 = (\pi^2/6)\delta^2$; with this initialization the Newton–Raphson method gives an estimate in half-dozen steps.

It is simple to show that $(\hat{\lambda}, \hat{\delta})$ are asymptotically normal with mean values λ and δ, variances $[1+(6/\pi^2)(1-\gamma)^2](\delta^2/n)$ and $(6/\pi^2)(\delta^2/n)$ and correlation $\{1+\pi^2/[6(1-\gamma)^2]\}^{-1/2}$.

The estimation of the parameters $(\lambda, \delta, \alpha)$ (α, a shape parameter) for Fréchet and Weibull distributions has some difficulties and will not be needed in the sequel. Note that although the fact that logarithmic transformations reduce Fréchet and Weibull distributions to Gumbel distribution is important theoretically, it is not useful for statistical decision unless we know the value of λ.

3. The unified form of maxima distributions

It was shown by von Mises and Jenkinson, see Gumbel (1958), that if we consider the distribution function

$$G(z|k) = \{\exp[-(1+kz)^{-1/k}]\},$$

if $1+kz > 0$ (with $G(z|0^+) = G(z|0^-) = \exp(-e^{-z}) = \Lambda(z)$),

with the natural continuation if $k < 0$ when $z > -1/k(>0)$ then $G(z|k) = 1$ and if $k > 0$ when $z < -1/k$ (< 0) then $G(z|k) = 0$, it is immediate that

$$G(z|k) = \Psi_{-1/k}(-1-kz) \quad \text{if } k < 0$$
$$G(z|0) = \Lambda(z) \quad \text{if } k = 0$$
and
$$G(z|k) = \Phi_{1/k}(1+kz) \quad \text{if } k > 0.$$

The decision that $k < 0$, $k = 0$ or $k > 0$ corresponds to the choice of one of the three extreme models.

Note that this choice is not a "separation" in the meaning of Cox (1961, 1962) because Gumbel distribution $\Lambda(z) = G(z|0^-) = G(z|0^+)$ can be obtained as limits of Weibull and Fréchet distributions when $\alpha \to +\infty$ ($\alpha = -1/k$ in the Weibull case, $\alpha = 1/k$ in the Fréchet case).

Thus we will develop a trilemma decision for the statistical choice of one of the univariate extreme models in the line of statistical testing, using the central role of Gumbel distribution ($k = 0$). The results, sketched here, are based on Tiago de Oliveira (1980b).

4. Statistical choice of univariate statistical models

Let $(z_1,\ldots,z_n)$ be a sample of IID random variables with distribution function $G(z|k)$.

The LMP test of $k = 0$ against $k > 0$ leads to the rejection region

$$\sum_1^n \frac{\partial \log G'(z_i|k)}{\partial k}\bigg|_{k=0} \geq a_n$$

and the LMP test of $k = 0$ against $k < 0$ leads to the rejection region

$$\sum_1^n \frac{\partial \log G'(z_i|k)}{\partial k}\bigg|_{k=0} \leq b_n.$$

As

$$G'(z|k) = \exp\{-(1+kz)^{-1/k}\}(1+kz)^{-1/k-1} \quad \text{if } 1+kz \geq 0$$
$$= 0 \quad \text{if } 1+kz \leq 0$$

we have

$$v(z) = \frac{\partial \log G'(z|k)}{\partial k}\bigg|_{k=0} = \frac{z^2}{2} - z - \frac{z^2}{2}e^{-z}.$$

Note that, denoting $\underset{\sim}{z} = \min(z_i) \leq \max(z_i) = \tilde{z}$, from the relation $1+kz \geq 0$ we get:

$$\underset{\sim}{z} \leq \tilde{z} \leq 0 \quad \text{implies} \quad k \leq -1/\underset{\sim}{z}$$
$$\underset{\sim}{z} \leq 0 \leq \tilde{z} \quad \text{implies} \quad -1/\tilde{z} \leq k \leq -1/\underset{\sim}{z}$$

and
$$0 \leq \underset{\sim}{z} \leq \tilde{z} \quad \text{implies} \quad k \geq -1/\tilde{z}$$

so that the hypothesis $k = 0$ is always admissible. We must then search in the neighbourhood of $k = 0$ which suggests the following.

The statistical choice of a univariate extreme model for the (reduced) sample $(z_1,\ldots,z_n)$ leads to the trilemma decision: choose b_n and a_n ($\geq b_n$) and if $\sum_1^n v(z_i) \leq b_n$ decide for Weibull distribution, if $b_n < \sum_1^n v(z_i) \leq a_n$ decide for Gumbel distribution and if $a_n < \sum_1^n v(z_i)$ decide for Fréchet distribution.

Denoting the decision for Weibull, Gumbel and Fréchet distributions by W, G and F, we have, when k is the true (unknown) value,

$$P_n(W|k) = \text{Prob}\left\{\sum_1^n v(z_i) \leq b_n | k\right\},$$

$$P_n(G|k) = \text{Prob}\left\{b_n < \sum_1^n v(z_i) \leq a_n | k\right\}$$

and
$$P_n(F|k) = \text{Prob}\left\{a_n < \sum_1^n v(z_i) | k\right\}.$$

6 DECISION AND MODELLING FOR EXTREMES

The probabilities of wrong decision (misclassification) are thus

$$1 - P_n(W|k) = P_n(G|k) + P_n(F|k) \quad \text{if } k < 0,$$
$$1 - P_n(G|0) = P_n(W|0) + P_n(F|0) \quad \text{if } k = 0$$

and
$$1 - P_n(F|k) = P_n(W|k) + P_n(G|k) \quad \text{if } k > 0.$$

As it is natural we will search an asymptotically optimal decision rule. Let $\mu(k)$ and $\sigma^2(k)$ denote the mean value and variance of $v(z)$ with respect to $G(z|k)$; note that $\mu(0) = 0$ and $\mu'(0) = \sigma^2(0) = 2.423\,6\bar{1}$ so that $\mu(k)$ is increasing around $k = 0$. It must be noted that $\mu(k)$ exists only for $-1 < k < \frac{1}{2}$ and $\sigma^2(k)$ exists only for $-1 < k < \frac{1}{4}$.

From the central limit theorem (in the region $-1 < k < \frac{1}{4}$) and Khintchine convergence of types theorem we see that

$$\text{Prob}\{W|k\} = \text{Prob}\left\{ \frac{\sum_{1}^{n} v(z_i) - n\mu(k)}{\sqrt{n}\sigma(k)} < \sqrt{n}\frac{b_n/n - \mu(k)}{\sigma(k)} \right\}$$

$$\to N\left(\lim \sqrt{n}\frac{b_n/n - \mu(k)}{\sigma(k)} \right),$$

$$\text{Prob}\{G|k\} = \text{Prob}\left\{ \sqrt{n}\frac{b_n/n - \mu(k)}{\sigma(k)} < \frac{\sum_{1}^{n} v(z_i) - n\mu(k)}{\sqrt{n}\sigma(k)} \leq \sqrt{n}\frac{a_n/n - \mu(k)}{\sigma(k)} \right\}$$

$$\to N\left(\lim \sqrt{n}\frac{a_n/n - \mu(k)}{\sigma(k)} \right) - N\left(\lim \sqrt{n}\frac{b_n/n - \mu(k)}{\sigma(k)} \right)$$

and

$$\text{Prob}\{F|k\} = 1 - \text{Prob}\left\{ \frac{\sum_{1}^{n} v(z_i) - n\mu(k)}{\sqrt{n}\sigma(k)} \leq \sqrt{n}\frac{a_n/n - \mu(k)}{\sigma(k)} \right\}$$

$$\to 1 - N\left(\lim \sqrt{n}\frac{a_n/n - \mu(k)}{\sigma(k)} \right),$$

where $N(x)$ denotes the standard normal distribution function.

As $\text{Prob}\{W|k\}$ must converge to zero if $k \geq 0$ and $\mu(0) = 0$ we must have $b_n/\sqrt{n} \to -\infty$ and for $k > 0$ (if $\mu(k) > 0$) we have, also, $\text{Prob}\{W|k\} \to 0$. Consider now $k < 0$: if $\mu(k) < 0$, as $\mu(k)$ is continuous around $k = 0$, to have $\text{Prob}\{W|k\} \to 1$ we must have $b_n/n \to 0$ (if not, let $Q < 0$ be one of the limits of b_n/n; under the corresponding subsequence, for k such that $Q < \mu(k) < 0$, we would have $\sqrt{n}[b_n/n - \mu(k)] \to 0$ and $\text{Prob}\{W|k\} \to 0$).

Thus the conditions on b_n are $b_n/\sqrt{n} \to -\infty$ and $b_n/n \to 0$. Similarly, from $\text{Prob}\{F|k\}$ we will get $a_n/\sqrt{n} \to +\infty$ and $a_n/n \to 0$.

Consider now the decision for Gumbel distribution. We have

$$\text{Prob}\{G|0\} \to N\left\{\lim \frac{a_n}{\sqrt{n\sigma(0)}}\right\} - N\left\{\lim \frac{b_n}{\sqrt{n\sigma(0)}}\right\}$$

and for $k \neq 0$ we get $\text{Prob}\{G|k\} \to 0$.

It is now natural to decide on the (possible) relations between b_n and a_n. The probabilities of (wrongly) deciding for Weibull or Fréchet distributions, if Gumbel distribution ($k = 0$) is true, are approximately $N[b_n/\sqrt{n\sigma(0)}]$ and $1 - N[a_n/\sqrt{n\sigma(0)}]$. If we impose the symmetry condition of (approximately) equal probabilities of error we get $a_n + b_n = 0$. We will, thus, take from now on $b_n = -a_n$, with $a_n/\sqrt{n} \to +\infty$ and $a_n/n \to 0$ ($a_n > 0$).

The approximate probability of deciding for Gumbel distribution if $k = 0$ is, thus,

$$N[a_n/\sqrt{n\sigma(0)}] - N[-a_n/\sqrt{n\sigma(0)}] = 2N[a_n/\sqrt{n\sigma(0)}] - 1,$$

so that the probability of a wrong decision is, approximately,

$$2\{1 - N[a_n/\sqrt{n\sigma(0)}]\}.$$

If we assume this probability to be γ_n ($\to 0$) we get

$$\frac{2}{\gamma_n}\{1 - N[a_n/\sqrt{n\sigma(0)}]\} = 1,$$

whose asymptotic solution is, as is well known,

$$\frac{a_n}{\sqrt{n\sigma(0)}} = \sqrt{(-2\log\gamma_n/2)} - \frac{\log(-\log\gamma_n/2) + \log 4\pi}{2\sqrt{(-2\log\gamma_n/2)}}$$

and it is immediate that $a_n/\sqrt{n} \to +\infty$. To impose the condition $a_n/n \to 0$ we must have $\sqrt{(-2\log\gamma_n/2)}/\sqrt{n}$ or $\log\gamma_n/n \to 0$. In particular, this condition is verified if we take $\gamma_n = 1/n$.

The idea of statistical choice is associated to the principle that the probability of wrong decision must converge to zero and the probability of correct decision must converge to one. Recall that although we can expect the results to be true for all k, the methodology presented is only true for k such that $-1 < k < \frac{1}{4}$ and when $\mu(k) < 0$ if $k < 0$ and $\mu(k) > 0$ if $k > 0$, a fact that was shown valid, at last, in a neighbourhood of $k = 0$.

The classical (LMP) asymptotic two-sided test of $k = 0$ against $k \neq 0$ could be dealt with by taking $a_n = -b_n = \sqrt{n\sigma(0)}(\chi_{1-\alpha/2})$ where χ_p is the p-quantile of the standard normal distribution. In that case we have only $a_n/n \to 0$ which guarantees $\text{Prob}\{W|0\} \to 0$ and $\text{Prob}\{F|0\} \to 0$; but $\text{Prob}\{G|0\} \to N(\chi_{1-\alpha/2}) - N(-\chi_{1-\alpha/2}) = 1 - \alpha$.

Let us now introduce the location (λ) and dispersion ($\delta > 0$) parameters. It is thus natural to use the "estimated" reduced values $\hat{z}_i = (x_i - \hat{\lambda})/\hat{\delta}$, where $\hat{\lambda}$

and $\hat{\delta}$ are the ML-estimators of λ and δ in the Gumbel case. With the δ-method (Tiago de Oliveira, 1980b) we see that $\Sigma_1^n v(\hat{z}_i)$ is asymptotically normal with mean value $n\hat{\mu}(k)$ and variance $n\hat{\sigma}^2(k)$ such that $\hat{\mu}(0) = 0$ and $\hat{\mu}'(0) = \hat{\sigma}^2(0) = 2.09797$, a reduction of 13.44% in the variance, as could be expected. We will decide for Weibull, Gumbel and Fréchet distributions according to $\Sigma_1^n v(\hat{z}_i) \leqslant -a_n$, $-a_n < \Sigma_1^n v(\hat{z}_i) \leqslant a_n$ and $a_n < \Sigma_1^n v(\hat{z}_i)$, where a_n was as defined before; the probabilities of a wrong decision converge to zero and of a correct decision converge to one, as before.

5. The natural bivariate extreme models

As is well known, if (X, Y) is a bivariate random pair with standard Gumbel margins ($\Lambda(x)$ and $\Lambda(y)$) its distribution function is

$$\text{Prob}\{X \leqslant x, Y \leqslant y\} = \exp\{-(e^{-x} + e^{-y})k(y-x)\},$$

where the dependence function $k(w)$ must satisfy some conditions, not stated here because they will not be used in the sequel; see Tiago de Oliveira (1975, 1980a) for details and for the formulae used below. Recall that the dependence function must verify the bounds $k_D(w) = \max(1, e^w)/(1 + e^w) \leqslant k(w) \leqslant 1$; $k_D(w)$ corresponds to the diagonal case, where we have $\text{Prob}\{Y = X\} = 1$, and $k(w) = 1$ represents the independence case. Let us once more stress that we are dealing with standard margins in this section.

Let (X_1, Y_1) be a pair of independent standard Gumbel random variables and consider the new (dependent) random pair (X, Y) defined as

$$X = \max(X_1 - a, Y_1 - b), \quad Y = \max(X_1 - c, Y_1 - d)$$

with $a, b, c, d \in R$.

It is immediate to show that X and Y are standard Gumbel margins if $e^{-a} + e^{-b} = 1$ and $e^{-c} + e^{-d} = 1$; thus we must always have $a, b, c, d \geqslant 0$.

Denoting $W_1 = Y_1 - X_1$ and $W = Y - X$, we see that

$$W = \max(0, W_1 + c - d) - \max(0, W_1 + a - b) + a - c$$

is a monotonic function of W_1 varying from $a-c$ to $b-d$. We will suppose from now on that $a-c \leqslant b-d$ (or $a+d \leqslant b+c$), interchanging, if necessary, (a, b, c, d) with (c, d, a, b) or X and Y. So we get $a-c \leqslant W \leqslant b-d$ and the random points (X, Y) are contained in a strip parallel to the first diagonal, imposing a strong stochastic relation between X and Y. From $e^{-a} + e^{-b} = 1 = e^{-c} + e^{-d}$ we get immediately $e^{-a}(1 + e^{a-b}) = e^{-c}(1 + e^{c-d})$ so that as $a+d \leqslant b+c$ we obtain $a-c \leqslant 0$. Similarly we obtain $b-d \geqslant 0$.

The origin then belongs to the strip. Denote now $a-c = -\alpha$, $b-d = \beta$.

We have, thus, two parameters (α, β) and the random pair (X, Y), concentrated in the strip $-\alpha \leq Y - X \leq \beta$, has the distribution function.

$$\text{Prob}\{X \leq x, Y \leq y\}$$
$$= \text{Prob}\{X_1 \leq \min(a+x, c+y), Y_1 \leq \min(b+x, d+y)\}$$
$$= \exp\{-(e^{-x} + e^{-y})k(y-x)\}$$

with

$$k(w|\alpha, \beta) = \frac{1}{1+e^w} \quad \text{if } w \leq -\alpha$$

$$= \frac{1 - e^{-\beta} + (1 - e^{-\alpha})e^{-w}}{(1 - e^{-\alpha-\beta})(1+e^{-w})} \quad \text{if } -\alpha \leq w \leq \beta$$

$$= \frac{1}{1+e^{-w}} \quad \text{if } \beta \leq w.$$

Recall that $\alpha, \beta \geq 0$ but the case $\alpha = \beta = 0$ corresponds to the diagonal case, as follows from the fact that $k(w|\alpha, \beta) = k_D(w)$ when $w \leq \alpha$ or $w \geq \beta$. The model is, evidently, identifiable.

The natural model contains the biextremal model when $\alpha = -\log \theta$, $\beta = +\infty$ ($0 \leq \theta \leq 1$) and its dual when $\alpha = +\infty$, $\beta = -\log \theta$ ($0 \leq \theta \leq 1$); independence corresponds to $\alpha, \beta \to +\infty$ and the diagonal case is obtained for $\alpha = \beta = 0$.

As the correlation coefficient between X and Y is

$$\rho = -\frac{6}{\pi^2} \int_{-\infty}^{+\infty} \log k(w) \, dw$$

and from the evident relation

$$1 = -\frac{6}{\pi^2} \int_{-\infty}^{+\infty} \log k_D(w) \, dw$$

we get

$$\rho(\alpha, \beta) = 1 - \frac{6}{\pi^2} \int_{-\infty}^{+\infty} \log \frac{k(w|\alpha, \beta)}{k_D(w)} dw$$

$$= 1 - \frac{6}{\pi^2} \int_{-\alpha}^{\beta} \log \frac{1 - e^{-\beta} + (1-e^{-\alpha})e^{-w}}{(1-e^{-\alpha-\beta}) \max(1, e^{-w})} dw$$

$$= 1 - \frac{6}{\pi^2} \left\{ -\frac{\alpha^2}{2} + (\alpha+\beta) \log \frac{e^\beta - 1}{e^\beta - e^{-\alpha}} + \int_{(1-e^{-\alpha})/(e^\beta-1)}^{(e^\alpha-1)/(1-e^{-\beta})} \frac{\log(1+t)}{t} dt \right\}.$$

It is easy to show that $\rho(\alpha, \beta) \to 1$ when $\alpha, \beta \to 0$ and $\rho(\alpha, \beta) \to 0$ when $\alpha, \beta \to +\infty$.

From the computation of $\rho(\alpha, \beta)$ we can obtain the linear regressions $y - \gamma = \rho(\alpha, \beta)(x - \gamma)$ and $x - \gamma = \rho(\beta, \alpha)(y - \gamma)$.

Let us now consider the general regression. As it is evident that the interchange of x and y implies the interchange of α and β it is sufficient to compute the regression line $\bar{y}(x|\alpha, \beta)$. It can be shown that

$$\bar{y}(x|\alpha, \beta) = x + \beta - \frac{e^\beta - 1}{e^\beta - e^{-\alpha}} \exp\left(\frac{1 - e^{-\alpha}}{e^\beta - e^{-\alpha}} e^{-x}\right) \int_{[(1-e^{-\alpha})/(e^\beta - e^{-\alpha})]e^{-x}}^{[e^\beta(e^\alpha - 1)/(e^\beta - e^{-\alpha})]e^{-x}} \frac{e^{-t}}{t} dt.$$

As was shown in Tiago de Oliveira (1974) the linear regression for the bi-extremal model and its dual has a very high efficiency; we can expect the same to apply generally.

It is very easy to show that $D(w|\alpha, \beta) = \text{Prob}\{Y - X \leq w\}$ has the expression

$$D(w|\alpha, \beta) = 0 \quad \text{if } w < -\alpha,$$
$$= \frac{1 - e^{-\beta}}{1 - e^{-\beta} + (1 - e^{-\alpha})e^{-w}} \quad \text{if } -\alpha \leq w < \beta,$$
$$= 1 \quad \text{if } \beta \leq w,$$

with jumps of $(e^\beta - 1)/(e^{\alpha + \beta} - 1)$ at $-\alpha$ and $(e^\alpha - 1)/(e^{\alpha + \beta} - 1)$ at β.

The estimation of the two location parameters α and β is easy to obtain when margins are standard Gumbel random variables. Denoting $w_i = y_i - x_i$, $\underline{w}_n = \min w_i$ and $\tilde{w}_n = \max w_i$ we have $-\alpha \leq \underline{w}_n \leq \tilde{w}_n \leq \beta$ so that the estimators of α and β can be taken as

$$\alpha_n^* = -\min(0, \underline{w}_n) \quad (<\alpha)$$
$$\beta_n^* = \max(0, \tilde{w}_n) \quad (<\beta).$$

The consistence of the estimators comes from the fact that $\text{Prob}\{\alpha_n^* = 0\} = [1 - D(0|\alpha, \beta)]^n$ and $\text{Prob}\{\beta_n^* = 0\} = D^n(0|\alpha, \beta)$ converge to zero as

$$0 < D(0|\alpha, \beta) = \frac{2e^\beta - 1 - e^{\beta - \alpha}}{2(e^\beta - e^{-\alpha})} < 1.$$

References

Berman, S. M. (1962). Limiting distributions of the maximum term in a sequence of dependent random variables, *Annals of Mathematical Statistics*, **33**, 894–908.

Berman, S. M. (1964). Limit theorems for the maximum term in stationary sequences, *Annals of Mathematical Statistics*, **35**, 502–516.

Cox, D. R. (1961). Tests of separate families of hypotheses. In *Proceedings of the 4th Berkeley Symposium on Mathematical Statistics and Probability*, Vol. 1, 105–123. University of California Press, Berkeley.

Cox, D. R. (1962). Further results on tests of separate families of hypotheses, *Journal of the Royal Statistical Society* B, **24**, 406–424.
Galambos, J. (1978). *The Asymptotic Theory of Extreme Order Statistics*. Wiley, New York.
Gnedenko, B. V. (1943). Sur la distribution limite du terme maximum d'une série aléatoire, *Annals of Mathematics*, **44**, 423–453.
Gumbel, E. J. (1958). *Statistics of Extremes*. Columbia University Press, New York.
Juncosa, M. (1949). On the distribution of the minimum in a sequence of mutually independent random variables, *Duke Mathematical Journal*, **16**, 609–618.
Leadbetter, M. R. (1974). On extreme values in stationary sequences, *Zeitschrift fur Warschein lichkeitstheorie und ihre Verwandte Gebiete*, **28**, 289–303.
Loynes, R. M. (1965). Extreme values in uniformly mixing stationary stochastic processes, *Annals of Mathematical Statistics*, **36**, 993–999.
Newell, G. F. (1964). Asymptotic extremes for m-dependent random variables, *Annals of Mathematical Statistics*, **35**, 1322–1325.
Tiago de Oliveira, J. (1974). Regression in the non-differentiable bivariate extreme models, *Journal of the American Statistical Association*, **69**, 816–818.
Tiago de Oliveira, J. (1975). Statistical decision for extremes, *Trabajor de Estadistica y Investigacion Operativa*, **XXVI**, 433–471.
Tiago de Oliveira, J. (1978). Approximate distributions for sequences of maxima. *Metron*, **XXXVI**, 3–21.
Tiago de Oliveira, J. (1980a). Bivariate extremes: foundations and statistics, *Multivariate Analysis V*, ed. P. R. Krishnaiah, 349–366. North-Holland, Amsterdam.
Tiago de Oliveira, J. (1980b). Statistical choice of univariate extreme models, *Statistical Distributions in Scientific Work*, vol. 6, C. Taillie *et al.* (eds), 367–387, Reidel, Dordrecht.
Turkman, K. F. (1979). On the Maxima of Stationary Sequences. Research Report No. 206/KFT 2, Manchester–Sheffield School of Probability and Statistics.
Watson, G. S. (1954). Extreme values in samples from m-dependent stationary stochastic processes, *Annals of Mathematical Statistics*, **25**, 798–800.

7

AN OVERVIEW OF TECHNIQUES OF DATA ANALYSIS, EMPHASIZING ITS EXPLORATORY ASPECTS

Colin L. Mallows and John W. Tukey

Abstract

This paper surveys the current technology of data analysis from a variety of viewpoints, though a complete account is not attempted. First, the model-data cycle is examined, and the classical idea that a model is "given" is commented upon. Several critical components of an appropriate general framework are discussed, and a taxonomy of styles of data analysis is developed. Several topical areas, such as regression, ANOVA, smoothing, and time series are treated briefly. Structural escalation, in which the data is used to modulate a weak *a priori* structure for the data into an apparently stronger *a posteriori* structure, is examined in detail, with reference to factor analysis, multi-dimensional scaling, cluster analysis and l'analyse des correspondances. A final section includes comments on computing, report-writing and on the current relevance of an earlier survey by one of the authors.

1. Introduction

We will not confine our attention to a specific slice of years, and we will find it necessary to outline philosophies and techniques

(i) as we feel they were *implicitly* as well as explicitly and
(ii) as we feel they now are or *ought* to be in the future.

Prepared in part in connection with research at Princeton University sponsored by the US Army Research Office (Durham).

With limited time and space, we shall have to focus on the basic structures and issues and emphasize existing or near-future accounts of specific areas as sources for further detail.

We do not attempt either a complete account of data analysis, which would require careful attention to how techniques are—or should be—chosen for application, or an account of the broader aspects of learning through collection and analysis of data, which would require both careful and extensive attention to both the choice of data to be collected and the techniques for its collection. All of these together with the assessment of data quality, are areas of major importance, left aside because of limitations of time and space.

We recognize that this survey is far from balanced. We have tried to give it enough different detailed slants to give some large-scale balance. Classical methods have certainly been unfairly slighted.

The role of data analysis

Until recently, the statistician's published view of his role—in relation to science and scientists—ran roughly as follows: Dramatic advances in science are rare, and can result either from fresh theoretical insight (relativity) or the observation of new phenomena (particle physics). Many more, though less striking advances are made by way of careful repeated experiments and detailed analysis of observations. In this latter activity the statistician's role is to aid in impartial assessment of the strength of the evidence for and against a particular model, and in favor of one or another particular value (or range of values) for the parameters in that model.

In recent years, partly because of explosions in the amount and variety of data that is available for analysis and in the computing power available to the analyst, a larger role that many have long carried out quietly and effectively has been more broadly recognized. Now the statistician is seen also as an explorer, searching for stable and meaningful patterns; he follows many trails and considers a great variety of possible models. Of course, the results of the analysis may be meaningless or trivial, but this danger is lessened both by deep understanding of the context and by sensible exploration.

As yet there is no prospect of being able to replace the data-analyst with a computer, though the analyst of massive data sets would be lost without the computer as an aid. The analyst must think about the possible interpretations and implications of his results, and be anxious to consult in depth with the subject-matter expert.

While we cannot expect that either the statistician or his client will make many dramatic breakthroughs, working together as closely cooperative colleagues they often make substantial advances in depth of understanding and precision of concept.

A. THE MODEL-DATA CYCLE

An adequate understanding of data analysis is impossible without a proper appreciation of the cyclic relation of model and data. Statistical theory has ordinarily taken the model (or, perhaps, models) as given and discussed the analysis of the data within this context. This may be (locally) correct, but is inevitably one-sided. Data has to be used to generate models for its analysis, and we have to learn to understand equally well this side of the cycle.

2. A history of divided responsibility: models are NOT "given"!

Most textbook discussions of statistics or data analysis take the model (in all its aspects) as given—almost as handed down from above, as the tablets of stone were to Moses. This may have become a custom because mathematical theory could be developed starting from unquestioned assumptions and such theory was easier to teach than data analysis.

The unquestioned existence of a model was usually claimed for both

(i) forms of functional dependences of one variable on another (each possibly multi-dimensional)—more precisely, usually, forms of dependences of a typical value of one variable on other variables—and
(ii) a statistical (or stochastic) model of how any one detailed dependence would be disturbed by perturbations, fluctuations and measurement error.

(Statisticians sometimes seem to overemphasize the second.) The collection of alternative straight lines

$$\{\eta_i = a + bx_i | \text{any } a \text{ and } b\}''$$

and the statistical model.

$$y_i - \eta_i \quad \text{independently from a Gaussian distribution with mean 0 and unknown } \sigma^2$$

is perhaps the simplest illuminating example.

Clearly such pictures of unquestionable models result from dividing the modeling from the data analysis in a quite unrealistic way. While there are situations where scientific or engineering theory suffices to suggest the details of such a model, they are very rare indeed. In the real world, most suggestions of models come from someone having looked at data—perhaps extensively and carefully, but perhaps not.

Being realistic about models; recognizing the

$$\left(\begin{array}{c}\text{model}\\ \text{data}\end{array}\right)$$

cycle as it exists, requires in particular that our view of models is made more realistic in another way we are about to describe.

3. Models as truth vs. models as leading cases

While relatively few of those who actually analyze data effectively were among those who really took seriously the philosophy of statistical models as truth, far too many lectures and books acted/spoke/wrote as if this were so—in particular that data follows a Gaussian distribution exactly.

When either physicists or mathematicians discuss the motion of a "mass point", or when the hard-round-sphere model for a gas is used, few of those concerned with mechanics in the real world have their ideas or insights diverted or disturbed. They know there are no mass points—and that molecules are not hard. But too many feel that one should not use the analysis of variance unless the residuals are demonstrably (nearly) Gaussian in distribution—an all too common error of over-reaction. This is only one instance where purism in statistics—the use of probabilistic models as truth, as a part of an image of data analysis formulated as on utopian inferential process—has both diverted energy and distorted understanding.

It is not easy to be as schizophrenic as a statistical theorist/analyst must be. To combine the apparent certainties of mathematical structures and theorems with the obvious uncertainties of limited data is not easy—and never will be.

We all need, however, to learn to use the consequences of precise assumptions as only leading cases—as every engineer has had to use the mechanics of point masses—and to expand the precise to an umbra in which we understand much of what is going on, and then to a penumbra in which we understand less. In almost the legal sense of those words, the mechanics of point masses is a "leading case". And the umbra around it is a domain of understanding.

We can live with statistical inference based upon Gaussian distributions as *one of several* "leading cases", but recent developments in robust/resistant techniques and theory should convince us that we are unlikely to be able to do well by keeping it—or any other narrow specification—as *the only* leading case. And we surely dare not think of it as truth.

Bland or flavored?

We need to distinguish between bland and highly-flavored statistical models. Bland models—whether rewarding or unrewarding per observation—do not allow statistical inference to be based upon minor or unrealistic aspects of the models. Highly-flavored models (such as those with distributions showing discontinuities in their densities), by contrast, do let us do such misleading things.

Thus, for problems of location—more precisely problems of slippage—the shape of the slash distribution, with density

$$A \frac{1-\exp(-z^2/2)}{z^2} dz$$

is bland, while the shape of the Cauchy distribution, with density

$$B \frac{1}{1+z^2} dz$$

is more highly flavored. These two distribution shapes are similar in the tails, behaving like

$$C \frac{dz}{z^2}$$

but they behave quite differently near $z = 0$, where the Cauchy is much more peaked (in a way very rarely shown by actual data, which usually supports Windsor's principle that "all observed distributions are Gaussian in the middle"). If we use the Cauchy as a leading case, then, the resulting analysis pays far too much attention to the very center of the observed distribution.

As general challenges to techniques, particularly techniques which must work reasonably against Gaussian challenges, sampling from both slash and Cauchy distributions is rather effective. As quantitative standards against which to seek high percent efficiency, the Cauchy model is dangerously misleading, while, so far as we know, the slash model is not. Thus the Cauchy is *not* a good leading case.

Today's needs

We need to develop, both vaguely and formally, a philosophy of models as leading cases and ways of describing the umbrae and penumbrae around them. We need to develop a more complete understanding of blandness and its antithesis, untrustworthy high-flavoredness.

4. What does the working data analyst most need to understand?

A thorough answer to this question would give much more structure to the subject of data analysis than we seem able to produce today. Some examples of areas where understanding ought to be increased can, however, be given, some on the subject-matter side (all data comes from some subject-matter) and some on the more formal side.

Data is not "given" either

A large and important collection of issues that we shall not attempt to survey in detail concerns data acquisition, data quality and data management. Often the main result of an analysis is the recognition that more data is needed; with growth of understanding comes sharper focus as to what new data should be sought out. Data analysis is just one phase of the larger cycle

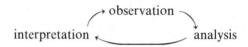

$$\text{interpretation} \leftarrow \text{observation} \rightarrow \text{analysis}$$

Hampel (1973, 1978) has demonstrated that even in disciplines with the highest standards, bad data is very common. Detailed study of the quality of data sources is an essential part of applied work. In exploratory analysis, use of robust/resistant techniques greatly expedites the search for the structures that are present in the majority of the data.

As soon as the amount of data to be studied, and/or the number of analyses to be attempted, is not small, simply keeping track of the data (including all the derived quantities) is a major problem. This is an area that is in very rapid development.

The measurement process

Data analysts need to understand more about the measurement processes through which their data comes. To know the name by which a column of figures is headed is far from being enough. It is often important, if not crucial, to understand how measurement is carried out, and what oddities are plausible. Surveyed distances are subject to different blunders if done with a 50-ft tape or a 100-ft one. The airplane pilot of weight 210 lb (95 kg) and height 29 in (74 cm) is understandable when one knows that the manual says that "double amputees are measured in a seated position". And many far less striking aspects of data can, and often do, fall into shape when the measurement process is understood. Meeting this requirement is only relatively easy.

"State of the art" possibilities

Data analysts need to understand, and take with a large grain of salt, which vaguely described mechanisms—which possible models—are reasonable. That prenatal nutrition *may* have an influence on "intelligence" at age 40, for example—something the naive statistician might not expect—can be relevant to certain sorts of data. Frequently, independence (between cases, or between variates) is very plausible, but the assumption should be challenged whenever data permits.

The stronger the qualitative understanding the data analyst can get of the subject-matter field from which his data comes, the better—just so long as he does not take it too seriously. It is only a few decades since the idea that a casual jingling of keys by a nearby psychologist might kill a rat would have seemed fantastic, yet just this happening led to the discovery of "audiogenic seizure".

Functional form and functional behavior

On the other hand, few of us know enough about the relationship of functional form and function behavior. How different are

$$a + b\log(y+100) \quad \text{and} \quad c + d\sqrt{y}$$

Over a wide range? Over a narrow range? For which narrow range will they be most similar? How do we see this? Can we solve similar questions about any pair of functional forms—any two parametrized classes of functions? Once we enter upon such questions we are led to even more interesting ones, such as: Which functional forms are more trustworthy in practice? Which less? Why?

To go a stage further, we doubt if any of us yet understands well enough the qualitative behavior of functions of two or more variables. Here it is hard to put the questions, yet the answers are very important.

5. An illustrative example

In 1973, the Governor of New Jersey, faced with a recommendation from his Department of Environmental Protection that to reduce ozone levels in the state to acceptable levels, the volume of motor vehicle emissions would have to be cut by two-thirds, and recognizing that this recommendation relied on only rudimentary understanding of the chemistry involved, asked Bell Laboratories to attack the scientific problem. A team of atmospheric chemists embarked on building a theoretical model of the atmosphere over New Jersey, while a team of statisticians collected and studied essentially all the available data, comprising eventually some 3 million measurements on some 30 chemical species and meteorological variables, taken at 60 sites over periods of up to 7 years.

One early result of the statistician's willingness to collect and organize the data was the observation that when plotted in time order, the values of the daily maxima of the ozone readings in the New York metropolitan area were just as high on Sundays as on weekdays. This was surprising, since it was known that in this area the volume of vehicular traffic is much lower on Sundays than on weekdays. This was an important clue, suggesting that the phenomenon under study was not simple and that much more detailed analysis was necessary.

A second striking observation, once the data had been organized and summarized appropriately, was that of all the 22 sites in New Jersey, the highest ozone readings were consistently those taken at Ancora, a rural community in southern New Jersey. Several explanations could be advanced for this.

One was that the Ancora readings were in error, due to poor calibration and maintenance of the recording equipment. Much effort went into a study of the reliability of the data.

A second hypothesis was that this result should have been expected; after all, a primary effect of the release of nitric oxide (NO) is to reduce the amount of ozone (O_3) by the reaction $NO + O_3 \rightarrow NO_2 + O_2$ so that ozone levels might be expected to be lower at urban sites than at neighboring rural ones. Painstaking study of the rates of literally hundreds of reactions, and their dependence on such factors as temperature and solar radiation was needed to properly assess the validity of this theory, thus dismissing it. Here close collaborations between statisticians and chemists was vital.

Yet a third hypothesis, which has turned out to be supported by other strong evidence, and so at this time is generally accepted, was that the ozone at Ancora resulted from emissions in Philadelphia, 23 miles away. This possibility was explored by collecting data on the wind direction at Philadelphia; a simple but ingenious graphical display of the (ozone, wind-direction) data (Cleveland and Kleiner, 1975) showed clearly that some association was present; later, exploration of data from southern New England gave strong support to the hypothesis of "transport" from metropolitan areas.

7 AN OVERVIEW OF TECHNIQUES OF DATA ANALYSIS   119

The crucial elements in both of these instances were

(i) a willingness to collect and study the data,
(ii) use of diagnostic techniques to show the unexpected,
(iii) an ability to recognize striking patterns (exceptionally high readings on Sundays, and at Ancora),
(iv) enough understanding of the context of the problem to enable these patterns to be recognized as potentially meaningful,
(v) avoidance of precipitate commitment to models of clearly inadequate complexity; much use was made of robust summaries and graphical displays and
(vi) energetic following-up of the clues obtained.

B. GENERAL CONSIDERATIONS

If we are to understand "the state of the art" adequately, we must do this in a suitable general context. The next twelve sections are devoted to developing some critical elements of such a context, often in a rather sketchy way.

6. Some basic concepts

We begin with brief comments on rather diverse issues.

Facing diverse challenges

At one time, formal judgment of a statistical method was unquestionably made in a single situation, usually in the situation where the technique did best. And most of these situations involved Gaussian, binomial or Poisson distributions. (One of us still recalls giving a seminar where a friendly listener spent most of the hour trying to persuade him that adopting a technique was equivalent to assuming that the hypothesis under which it did best was true!)

Today many of us—and soon, we believe, most of us—take most seriously the use of robust/resistant techniques—techniques chosen to do very well in the face of diverse challenges, rather than surpassingly well in the face of only a single challenge.

Fewer of us have yet recognized the need of the future to be still further diverse in the challenges our techniques will do well against. The writers, however, are firmly convinced that this will prove to be a major path of further advance. (Other advances will involve *combining* things we already know how to do separately, as illustrated in Section 16, or by providing critical data analysis, see Section 7, for the results of multi-dimensionally-guided structural escalation, see Section 29ff)

Variance, and other measures of closeness

We have become accustomed to judge the quality of point estimates—and their generalizations—in terms of variance (or, sometimes, mean square error). We have learned (e.g. Andrews *et al.*, 1972) to look at other measures as well, in the hope—more often justified than might seem reasonable—that they will generally tend to agree. We need to recognize that variance *is* a good measure when there is agreement among measures, but that it may give too much importance to rare large deviations when measures disagree.

Thus, variance is not a primary standard, not a basic criterion. It is often, but not always, a useful auxiliary standard.

Choice of technique, how guided?

As we suggest slightly more fully in Section 7, the choice of techniques is neither always to be guided by empirical trial on real data nor always by how it performs within specific probability structures. And the latter evaluation may sometimes better come about by experimental sampling or Monte Carlo, sometimes by mathematics.

Borrowing strength

One of the fundamentals of the analysis of variance—and of all the probability models that end with "$\cdots + \varepsilon_{ijk}$", and include "where ε_{ijk} has variance σ^2"—is *borrowing strength* by using information that might be irrelevant, as when the comparison of treatments C and K is given a standard error or a confidence interval based on the sizes of the residuals of *all* treatments, not just those for treatments C and K. Knowing when to borrow and when not to borrow is one of the key aspects of statistical practice.

We are only beginning to recognize the many roles of borrowing strength. We need to do this more rapidly, more widely and in more diverse situations.

Practical and fundamental measurement

Campbell (1919, 1957) wrote very carefully about fundamental measurement, and its role in the discovery of "The Laws of Physics". Unfortunately, no author of equal stature has written about the role of practical measurement in dealing with most of the questions of science and technology that concern the world.

The idea that "The Fundamental Laws" were the proper target of scientific work, seems to have been accepted uncritically by many behavioral scientists, with the result that many Ph.D. theses were thought to be attempting to establish a fundamental law. It is not recorded that Hooke got a Ph.D. for "ut tensio, sic vis". Who knows of a physical science Ph.D. thesis that did establish a fundamental law? As the idea of fundamental laws were transferred to new fields, it seems to have been far too easy to forget just how hard it is to establish a fundamental law.

Let us admit that perhaps *one-tenth* of *one percent* of science's quantitative effort is devoted to the study of fundamental laws. (Here we include with science such fields of technology as engineering, agriculture, medicine and education.) That means that perhaps a few tenths of one percent of science's quantitative work needs fundamental measurement. The rest will do quite well, thank you, with practical measurement of an appropriate kind. And will call for correspondingly practical techniques of data analysis.

To say we require fundamental measurement on an interval scale before using an arithmetic mean may perhaps be correct for that one-tenth of one percent of the time when we are concerned with fundamental measurement. Yet, who would deny the arithmetic mean to the climatologist—purely as a matter of principle—because most temperatures are measured on a practical scale or because interval-scale temperatures have been available for only about a century? Practical measurement is by far the most used and the most analyzed. We ought, almost all the time, to choose our data analytic tools for their effectiveness and their robustness and not in terms of practical measurement vs. fundamental measurement. If we can be practical, in all sense, 999 times in 1000, we will have done well indeed.

Practical measurement should, however, most usually be measurement (part of a significant figure, at least) and not just directionality (sign of a difference). We hate to speak or write of "measurement" or say that something has been "measured" just because a few tests of significance (we hope very strongly they are indeed tests of directionality) have given positive results. To a chemist or physicist, "measured" will usually mean at least one significant figure—often two, three or several—which means, roughly, a standard error no more than $1/12$ ($1/120$, $1/1200$ etc.) of the measured value. "Of established direction", say at 5%, need not mean "measured".

Multiplicity, explicit or implicit

Looking at the seemingly largest (or best) is something required by good science or good technology. Adapting our critical procedures to take account of the fact we have done this is essential, as when H appears highest, but K lowest, out of many treatments—so we apply our critical data analysis to $H - K$.

Many are now used to simultaneous confidence intervals as well as individual ones—all should become used to both. All should recognize, too, that it is not clear that we ought always work to a "5% simultaneous", for example, the "LSD" which corresponds to "5% individual" seems to have served a useful function for years in, for example, the *American Potato Journal*. What is clear is that we owe it to ourselves

(i) to be clear about the differences between "simultaneous" and "individual", and
(ii) to know which is involved in any *P*-values, confidence limits etc. that we quote.

Such problems of selection out of a quite specific multiplicity are not always easy, but they are much simpler than those of selection out of a hazy collection of alternatives. The latter arises every time a single set of data is first adequately explored and then critically (in the sense of the next section) analyzed (often best in a way suggested by the exploration). The amount of multiplicity involved—the diversity of the plausibly possible results of exploration—is often very hard to assess. (See also Section 24)

Fitting, decomposition, structuring

Autonomic data analysis (analysis by application of an algorithm) takes on a variety of forms. Among them, three deserve to be single out here:

(i) Fitting, where observed values are separated into an (incompletely) described part—the fit and, what is not yet described, the residuals.
(ii) Decomposition, where observed values are separated into (usually three or more) parts such that at least two parts are described separately.
(iii) Structuring, in which a structure not previously obvious is applied to (or recognized in) a data set—usually by structural escalation, in which an obvious simpler structure is combined with the disposition of the data points relative to one another. (See Sections 8 and 29ff.)

All are important, each has its role.

7. Styles of data analysis

EDA, ADAP, and DDAP; CDA, OCDA, SCDA, and CCDA

It will help us to distinguish several flavors of data analysis, beginning with:

(i) EDA (exploratory data analysis)—detective work on data, directed to the question "what seems to be going on?"
(ii) CDA (critical or confirmatory data analysis)—responsive to the question

7 AN OVERVIEW OF TECHNIQUES OF DATA ANALYSIS

of "what seems established beyond a suitable doubt" (moderate, reasonable or extreme, as may be appropriate) as exemplified by standard errors, tests of significance (specifically, tests of directionality), and confidence statements.

Here the term "critical" is of general application, referring to assessments of stability of whatever quality—quality combining thoroughness and usefulness. On the one hand, quality improves (assuming equal trustworthiness) as we move down through a list of styles, such as

(i) standard errors,
(ii) tests of significance,
(iii) tests of directionality and
(iv) confidence statements.

On the other hand, quality also improves as we allow more carefully for multiplicity (e.g. by making statements of simultaneous rather than individual confidence) or for serendipity in the form of "capitalizing on chance" (e.g. by confirming on data not used to suggest the form of analysis). "Confirmatory" ought probably be confined to instances where the quality is high.

We can usefully subdivide each of these in at least two, thus for EDA:

(i) ADAP (autonomic data analysis processes)—the carrying through of predecided-upon or self-steering processes whose outcome is an analysis (not directly a guide to further steps).
(ii) DDAP (diagnostic data analysis processes)—processes whose purpose is to suggest other things to try or look into.

(This subdivision usually divides processes from one another, since one process rarely, if ever, serves both purposes.)

(i) OCDA (overlapping critical data analysis)—in which data already explored (in such a way that it *might* have appeared for critical data analysis in some other shape) is subjected to critical data analysis.
(ii) SCDA (separated critical (or simple confirmatory) data analysis)—in which new data, analyzed in a way suggested by other data, is subjected to critical data analysis (such analysis is truly confirmatory).
(iii) CCDA (careful confirmatory data analysis) in which design of data gathering, prechoice of what is to be assessed, and the most careful methods of critical data analysis are combined into the most trustworthy kind of gathering and assessment of evidence that we know. (Particularly needed in questions of broad public interest, like the efficacy of an important new drug or the efficacy of cloud seeding. A special case of SCDA.)

(This subdivision mainly refers to the situations in which processes are used. A single process may well be used in any or all of the three subdivisions.)

So far as we can describe it today, exploratory data analysis consists of alternating phases of autonomic data analysis and diagnostic data analysis, repeated as many times as may seem appropriate. The most frequent human role in exploratory data analysis is the interpretation of the diagnostics and the selection of the next autonomic analysis. As a consequence, as many diagnostic procedures as possible should end in *pictures*, since

(i) pictures communicate well to humans, and, indeed,
(ii) pictures are by far the best way to communicate the unexpected.

One aspect likely to be important in the making of usefully diagnostic pictures is conveniently tagged as *cooperative diversity*, in which different parts of the data are asked to provide values that may agree—indicating that our analysis is satisfactory, at least in this one aspect—or disagree—indicating the need for an enhanced analysis and, we hope, pointing diagnostically toward a preferred direction of enhancement.

A role for both

If either critical or careful confirmatory data analysis is going to continue to help us with important questions, as it has done throughout the twentieth century, its techniques will have to be applied to the results of *well-chosen* experiments or data collections. While theoretical insight or parallel knowledge will sometimes guide us in our well-considered choices, we will often have to be guided by what has come from the analysis of *earlier* experiments or data collections. These results will, ordinarily, have been hints, or appearances not yet beyond a reasonable doubt. They will have been provided by exploratory analysis, not critical/confirmatory analysis.

Most previous discussions have not clearly separated from one another either the components of the first pair (ADAP and DDAP) or those of the second triple (OCDA, SCDA and CCDA). As we learn to make more thoroughly—or, conceivably but improbably to give up making—these secondary distinctions, we will be likely to change the way in which we think about them.

We can be sure, however, that there will be a continuing need for *both* exploratory on the one hand, *and* critical/confirmatory on the other. We need *both* today, we will need both *tomorrow*.

How many places for probability?

Mathematical statistics—and critical/confirmatory analysis—have always made major use of probability ideas and probability mathematics. Over the decades we have gradually learned to do good jobs with fewer assumptions—today's robust/resistant techniques are our furthest step in this direction. Combined with an idea fundamental to Student's thinking—and less pervasive of Fisher's—that *our statistical inferences can only go to what was sampled* (*and not, usually, all the way to where we want to go*), the use of *robust/resistant techniques* offers us a framework for such analysis that is far less arbitrary than any we have struggled with in the past.

But what about exploratory analysis? Having probability ideas in the back of our mind while planning how to do exploratory data analysis seems to us to be reasonable. (We should not like to go as far as Benzecri (Benzecri *et al.*, 1973, p. 3) and start a principle with "Statistique n'est pas probabilite".)

Our middle position would not keep us from either

(i) setting up particular procedure for exploratory data analysis without any conscious attention to probability ideas or structures, or
(ii) assessing a proposed EDA procedure, not only by testing it on real data, but also by testing it—by mathematics, experimental sampling, or Monte Carlo techniques, as may seem easiest—against the synthetic data corresponding to some probability structure.

It is hard to have one's cake and eat it, too. But it is not really hard—only psychologically hard—to use probability for some purposes, but lay it aside for others. More of us must learn to do just this.

Qualities of result

We must be ready, then, to make proper use of quite different kinds or qualities of result:

(i) EDA, combining ADAP and DDAP, will tell us about appearances, sometimes offering some guidance as to their stability.
(ii) Critical data analysis, OCDA or SCDA, will offer us numerical insight into more or less limited aspects of the stability of our results—whether by standard errors or by confidence statements—aspects that may or may not take account of one or both of (a) diversified exploration and (b) possibilities of multiplicity in the analysis of the data before us.
(iii) Careful confirmatory data analysis, CCDA, will offer us the most reliable results we know how to gather.

Each can be important in its own place and in its own right, we need to become accustomed to the use of all.

The classical paradigm is "exploration takes place on one set of data, but confirmation on another"—something that would happen generally in an overideal world (were there one) and does happen very specially in those cases where careful confirmation is required. However, we will have to learn that there will be many instances in the real world where extensive exploration of a body of data will be followed by critical analysis along the lines that exploration showed to be most promising. As was indicated in the last section, we cannot value such analysis as highly as those that are separately confirmatory—those in which *new* data was used for confirmation. But equally, we cannot realistically assume that such analyses can obviously be avoided. We have to prepare ourselves for critical analyses that differ in quality in ways other than the numerical values of p that are calculated for them and the style of result (standard error? confidence statement?). And it may *not* be—though of course it may be—that what we ought to desire, in some particular situation, is the highest quality of critical analysis.

8. Structural escalation in general

One of the most noticeable trends in data analysis over recent decades is the expansion and broadening of procedures of *structural escalation* in which an apparently weaker *a priori* structure for the data is combined with the actual distribution of datasets to generate an apparently stronger *a posteriori* structure.

Classical examples

The simplest, and most ancient, examples are, presumably, the decisions (a) to fit a straight line to a group of (x, y) pairs and (b) to make a row-plus-column fit to a row-by-column array of responses. In each of these situations there have been, for some time, more or less familiar *diagnostic* procedures which can rather often suggest the need (or perhaps the absence of a need) for re-expressions of the variables, or modifications in the structure of the fit. Positive diagnoses from such procedures should—and usually do—lead to modified analyses.

Three steps

One or more elements of these linked steps,

(i) autonomic fit,
(ii) diagnosis and
(iii) enhanced refit (diagnosis-guided),

are available for a wider and wider variety of *a priori* structures; accordingly, our more detailed discussion, Sections 27ff, will be organized according to these *a priori* structures.

We also need to address the question of where the various steps of structural escalation stand in autonomic-diagnostic-critical-confirmatory terms, and, more generally, the broader aspects of the relations of such steps to probability models and stochastic modeling. This is an area of wide divergence: some seem to believe that data analytic steps cannot and should not be considered in the absence of a rather narrow probability model. Others seem to feel that "l'analyse de données" should be conducted almost free of any stochastic modeling. Our own views are intermediate, and we shall try to support this position by both brief general discussion and illustration in connection with specific procedures.

Autonomic fits

So far as we can see, the selection of classes of "autonomic fit" procedures has to be an empirical matter—those classes that seem to work well enough in a wide enough variety of situations will be routinely tried, either almost always or always. (Since the cost of trying uselessly is often far less than the cost of missing by not trying, it may well be worthwhile to try *always*.) Mathematical neatness, or relationship to *mathematically* interesting structure (or theorems), can hardly be given any weight, while experience in the value of results, principally as expressed by their interpretability and clue-generating power, has to be taken most seriously.

Diagnosis

The diagnostic steps are a more complex problem. Their evaluation is too likely to be approached in an unbalanced way from one of two quite different directions:

(i) do the data suggest (by not reaching some prechosen level of significance) that we are not forced to go further or
(ii) even if we are predisposed to go further, does it seem hopeless to do so?

Clearly the stochastic structure is more important—and the critical P-values are more extreme—when we approach from the first of these directions. (The adequacy of P-value guidance can be disputed for approaches from the second direction.)

We have not thought hard enough about how diagnostic processes should be evaluated. It is clear that we need to consider (a) what appearances that would interest us are moderately likely to occur, (b) what appearances that would interest us seem possible, (c) how well each particular DDAP seems likely to function vs. each of these and (d) how well the DDAP seems likely to do in detecting the unexpected. Should we go further? How should we try to do each of these things?

Enhanced fitting

The conversion of positive diagnoses into modified or generalized fits—into enhanced fits—has a third flavor. Subject-matter knowledge and general experience necessarily play major roles in this step—there is real room for doubt about the role of probability modeling.

Retrodiagnosis

So far, this discussion has omitted the particular step (of critical data analysis) which to many typifies statistics (usually statistics in the framework of confirmatory data analysis) namely *retrodiagnosis* as expressed in the form

> can I use this body of data alone to show the importance of necessity of including this particular aspect of the structure in a description of the data (through what is usually called a *test of significance*)?

Retrodiagnosis is important, but not overriding. Its apparent answers can easily be overridden by considerations outside the data at hand. Thus:

(i) If we have measured heights of 5 men and 5 women and found "no significant difference" it would be foolish to "accept the hypothesis" that men and women have the same average height.
(ii) If we had found $A > B$ at 1%, it would be foolish to believe that we know $A > B$ under circumstances where this comparison is the most significant of 1000, 100 or even 10 studies of a similar type and quality. (After all, it was Fisher (1926) who said "A scientific fact should be regarded as experimentally established only if a properly designed experiment *rarely fails* to give this (5%) level of significance.")

In emphasizing (a) autonomic fit, (b) diagnosis and (c) guided refit, we do not intend to diminish the role of retrodiagnosis, only to balance our thoughts and broaden our options.

An important aspect of retrodiagnosis is shown when we decide to keep some parts of the structure suggested by an autonomic fit, while discarding (or setting aside) other parts. This possibility makes it quite likely that autonomic fits should be designed to propose too much structure.

Our first classical example, the straight-line with non-zero slope is, of course, hard to divide into parts where we can keep one part and discard the other. The second classical example of quite the opposite character: not only can we easily keep row differences while forgetting column differences (or vice versa) but we can keep the differences *among* subgroups of rows [columns] while discarding differences *within* the same subgroups. From the point of view of separability, the second example is much more nearly typical.

Confusion

One of the main reasons for confusion among the roles of diagnosis, enhanced refit, and retrodiagnosis is the rather natural, temporary pattern in which diagnosis seems to be most simply (possibly even, most effectively) done by the conduct of one or more *autonomic enhanced fits* followed by retrodiagnosis of the additional aspects of these enhanced fits. (Once the retrodiagnosis is automated, the whole process may become autonomic.)

Our first classical example, where the fitting of a straight line was for so long felt to be naturally followed by the fit of a quadratic—and where the additional reduction in sum of squares of deviations, as we pass from linear fit to quadratic fit, was for so long taken as *the* way to diagnose "curvature", whether near to a quadratic or far from it—offers a fine example of how easily such confusion can arise.

9. Arithmetic and graphics

The role of *display* in data analysis is of extreme importance. The majority of all quite unexpected results are detected from display. Technology has changed since Karl Pearson's earliest days, when graphical mechanics was important as a computational device. Arithmetic gets yearly cheaper, so one has to work hard to spend even as much on arithmetic as it would cost to enter the data by hand. Using arithmetic to drive display becomes easier, but not at as fast a rate.

Today arithmetic is wisely used freely and generously in the service of display—in preparing the numbers that guide the display. And graphic techniques are used to communicate to the persons involved—heavily to communicate about results, but also to help a person think about what other things might be done and how the results might be interpreted. Sometimes this will be on the scratch pad, but more often on the display itself.

The importance of graphics for reporting the result of DDAP was stressed in the last section, today its importance for reporting the results of ADAP is growing rapidly (e.g. the two-way plot (Tukey, 1977, Chapter 10)), especially in more complex situations like multiple comparisons (for notched box plots see Larsen *et al.* (1978)).

Graphics can often suggest what arithmetic might well be done next. In this sense, there is an arithmetic-graphics cycle, but there ought to be no doubt that the graphics is the master and the arithmetic the servant. Richard Hamming's famous aphorism "The purpose of computing is insight, not numbers" applies almost equally well with "computing" replaced by "data analysis".

Thus a balanced account of data analysis would spend far more space on graphics than we shall be able to. We must do what we can by reference, particularly to work in progress by Chambers et al. (1981) and by Wainer and Thissen (1981), and, for a display-more-for-display's-sake mode, to the books by Bertin (1973, 1977, 1980) and Monkhouse and Wilkinson (1963).

10. Interactive arithmetic and interactive graphics

Exploratory data-analysis is necessarily an iterative process, in which one performs many tentative analyses, following up promising leads, discarding others, and searching always for strong, stable and meaningful patterns in the data. Much of this can be done with pencil, paper and patience. (Most of the techniques in EDA (Tukey, 1977) were developed in this mode.) An inexpensive hand calculator, with four arithmetic operations and no more than one memory location, can be very helpful in reducing the arithmetic tedium. A programmable hand-held calculator can, of course, be even more helpful. However, when the number of items rises much above 100, more extensive computerization of the effort becomes almost essential.

One of the most pervasive attributes of classical statistical methodology is the attempt to condense the data to its barest essentials—to a low-dimensional sufficient statistic, to a single posterior distribution, to an "analysis of variance table" (emphasizing mean squares) or to an optimum decision. These attempts are entirely appropriate when a probability model for the data is really firmly known, when a personalistic attitude is appropriate and when the consequences of the possible decisions can be assessed with precision.

However in exploratory work none of these conditions holds, and one must expect the analysis to generate substantial amounts of output. Consider for example a simple data-matrix containing values of p variates for each of n sample units. If p is 10, 2 of which deserve attention as responses, and the other 8 are roughly equally usable as carriers of regression, and n is 200, simply staring at the matrix will rapidly bring on tedium—insight will come but slowly, if at all.

An exploratory analysis of such a matrix, using only simple computer programs, might begin, as follows.

7 AN OVERVIEW OF TECHNIQUES OF DATA ANALYSIS

Examination of single variates

(i) The ten marginal distributions (we advocate stem-and-leaf diagrams) are computed and displayed. (Some would also ask for letter-value displays.)
(ii) Gross outliers are identified and removed (their values are set to "absent", their identities are specially stored with their values) for later study.
(iii) Plausible simple re-expressions (e.g. $\log x$ or $x^{k/2}$, for small integer k) are noted for possible later use. (Less simple ones might also be looked for.)

Examination of pairs of variates

(iv) Each variate is cut at (or near) its median and its hinges (quartiles), and the 45 4×4 frequency tables are computed and displayed, one for each pair of variates. Cells containing the conditional medians are marked in each table and any tables not reasonably derivable from a bivariate Gaussian distribution (in re-expressed variates) are repeated with + or − deviating cells identified. (Corner-sum values (Olmstead and Tukey, 1947) might well be calculated also, since looking at the extremes separately can make us more content to have only quartered each variate.)
(v) Some of these tables are manually identified as being "interesting", some perhaps by exhibiting strong dependence, or weak dependence in an unexpected direction, or absence of dependence where this was expected.
(vi) Selected pairs of variates are displayed as scatter-plots with smooth middle traces (Tukey, 1977). (Some might like all pairs or raw and re-expressed variables, and/or, perhaps, hinge or eighth traces.) If there are many points that number will be reduced by aggregation or sharpening or both (Tukey and Tukey, 1981). (A return to (iv) using (fit, residual) may be appropriate.)

Automated global examination of variates (less elementary)

A reasonable autonomic procedure for dealing with the relations of the 2 responses to the 8 carriers is badly needed. If we had it, it should be the third phase. We have pieces out of which such a program could be assembled, but our insight into the manifold possibilities that could occur with different bodies of data is not yet adequate to let us plan a definitive program. One of us would expect to include applying some form of component finding to the 8 carriers, regressing each y either on components or on both components and carriers, taking appropriate steps to shed terms with small coefficients, and examining scatter diagrams (like vi), both for the two y-residuals and for selected combinations of y-residuals and carriers or components.

At this stage, particularly while the third phase is not available, more outliers might be identified; tentative linear dependencies suggested; and new variates formed by combining pairs or triads of the original ones. Further stages of analysis could entail multiple regression, cluster analysis, principal component analysis etc.

The important points here are that at each stage the analyst chooses among several options; his strategy is guided by fairly large amounts of output; at none of the stages described above is any final decision made; quite possibly none of the displays obtained to the point described will appear in the final report.

Several excellent examples of exploratory analysis of relatively nicely structured cases of data are given in the book *Fitting Equations to Data* by Daniel and Wood (1971). (See especially Chapter 5.) Here a computer was used in batch mode, with repeated runs of regression program that produces a considerable volume of output—numerical lists of fitted values and residuals, various plots of the residuals, and parameter estimates, with their standard errors and covariance matrix. One advantage of this mode of operation is that it gives the analyst time to think between runs. However, on balance, and especially in the early stages of an analysis, more flexibility and faster turn-around are often desirable. Several more-or-less interactive computing systems are now available offering varying amounts of flexibility and convenience.

11. Tracking

There are two very different purposes for which a trace of the operations applied to the data could be of great value.

(i) A trace of what was done for exploration is particularly valuable, as J. L. Dolby has pointed out to us, when it includes comments as to why certain avenues were discontinued as unpromising.
(ii) A trace of what was done for final estimation or criticism has its greatest value in providing an adequate record of just how recorded results and summary or adjusted values made available to others were obtained.

Far too little attention has been given to such tracking.

Some argue that easier tracking is a strong argument for batch calculation rather than interactive calculation, even for exploration. Dolby has also convinced us that programming an interactive system to give an adequate track of explorations is feasible. If, as we argue above and in Section 46, interaction should take place in intermediate-sized steps, both the importance and the feasibility of tracking are enhanced,

12. How far can we automate the analyst?

We are just beginning to face this important question. We should look to such possibilities with anticipation, not fear. As Edgar Anderson once told one of us, the way to get new ideas is to tell the world about old ones. Equally, the way for data analysts to spend their time on more subtle and interesting questions is for the more routine matters to be automated.

In doing this, we must be realistic about what a data analyst really does. This implies, to us at least, that more than half of what will be automated will be diagnostic rather than therapeutic. One function that it would be most valuable to automate would be the displaying of a list of cautions and questions about a data set. We don't know how to do this very well, but mainly because we haven't tried hard enough. (Even the editing-type questions are not easy to plan for.)

13. General strategies

As we look to the past, the present and the near future, we can see a sequence of eras (with two separating branches) each of which is roughly characterized by a general strategy. While any particular era will have gone on at different dates in different areas of application—and may have been omitted in some— we can usefully mark them off by events (which often focus on people). A trial sequence would be

Era 0 (ultraclassical). (How was this, really?)

Unconformity 1 (Galton, Karl Pearson). We ought to gather enough data to be sure of our results (K.P.), as for example the distribution of sizes of crabs.

Era 2 (middle classical). Pearson distributions, χ^2 goodness of fit, standard errors, bivariate Gaussian distribution.

Unconformity 3 (Student, Fisher). Methods for presumably precise inference from small samples.

>*Branch Unconformity 3* (Egon Pearson and Student)*. Earliest inquiries into effects of failure of assumptions.

Era 4 (neoclassical). Accept response (and circumstances) in whatever numerical form is given—expanding in polynomials if needed. Use rigid specifications without a qualm. Weight only for precision. Development of formalisms: Fisher, Neyman–Pearson, Wald, Stein, naive Bayes.

Unconformity 5 (Campbell vs. Stevens). Emphasis on fundamental measurement by Campbell, with assertion that physical measurement was the only worthwhile kind. Introduction of scale types by Stevens—softer measurement can have distinctive properties, too. Attempts to make all data analysis "fundamental" by restricting statistical procedures to those invariant under changes corresponding to the scale types of the data. (This was supported, at least in some fields of application, by the introduction of "non-parametric"— that is, distribution-free—methods, sparked by Fisher and Frank Wilcoxon.)

Era 6. Do everything—well almost everything—in a non-parametric way; guard against 5% not being 5%; pay little attention to efficiency.

Branch Unconformity 7* (*de Finetti, Good, Savage, Lindley, Tversky, Dempster, Shafer*). Emphasize the analyst's perceptions—explore alternatives to naive Bayesianism.

Era 8*. Neo-Bayesian data analysis.

Unconformity 7 (*Tukey, Huber*). Try to make analyses efficient over a wider and wider range of alternative situations while paying some attention to the error rate.

Era 8 (*1980 model*). Do everything twice, once neoclassically and once robust-resistant, thinking hard whenever the answers are seriously different.

Unconformity 9. Learn to use the data to guide both re-expression and choice of functional form. Recognize that functional forms are more often conveniences than fundamental facts.

Era 10 (*Coming soon*). Use of data, both at hand and in parallel, to guide both re-expression and functional form choice more and more effectively. Construction of parallel neoclassical and robust-resistant analyses, both based upon re-expressed data and data-guided functional forms. Recognition of the rarity of data deserving fundamental measurement and suitably restricted analysis. Emphasis on recognizing and understanding the role of measurement error.

14. Unusualness

Fitting a functional form is always a matter of INCOMPLETE description. Our task is not over once we have found and discussed a fit. It is urgent that we also find and discuss the residuals.

In looking at the residuals, which often are the custodians of all information about how it might be well to alter the model, we have to keep both aspects of the model,

(1) functional behavior, and
(2) stochastic character,

in mind. Indeed the impact of the residuals on the functional behavior of the model ought usually to be greater than their impact on its stochastic character, if only because we will have—if we are wise or lucky—learned how to describe the model and choose the character of its fitting procedures in such a way as to be relatively unspecific about its stochastic character.

7 AN OVERVIEW OF TECHNIQUES OF DATA ANALYSIS 135

We have to be careful, in interpreting the appearance of the residuals, to understand as best we can the extent to which this appearance might reasonably be pure chance. Cox and Snell (1968) show how this can be done in a general class of models. However, we cannot let ease of doing this be a sole guide to providing us with new definitions of residuals unrelated to their meaages about functional behavior. Reformulating the residuals into a space orthogonal to the fit, thus creating "BLUS residuals", as suggested by Theil (1965, 1968), and others, can lead to elegant distributional results under the Gaussian null-hypothesis, but so far only at the cost of destroying the 1–1 relationship between residuals and data-sets that still seems essential both for interpretation and for improving the assumed functional behavior.

As we pass to less balanced situations (e.g. regression rather than simple comparisons or balanced anova) we are likely to find it desirable to supplement functional-behavior directed residuals with both leverage values (Hoaglin and Welsch, 1978; Andrews and Pregibon, 1978) (that tell how hard that residual pulled on our fit) and adjusted residuals (modified to allow for different effects of fitting at different points, e.g. Anscombe and Tukey, 1963).

15. Shrinkage

Data comes to us from measurement. No measurement is perfect—and most real phenomena, even if measured perfectly, do not fit our models exactly. Thus the collective behavior of our residuals (often better of our adjusted residuals) can tell us much about the stochastic character of what really happened.

When we are interested in our measurements collectively, as when

(1) we are concerned with a true distribution underlying the measured values (especially in situations like distributions of properties of stars, where there is no doubt that there are true values),
(2) we are concerned with the most unusual (largest, smallest etc.) of the measured values and
(3) we are trying to assess coefficients that appear in the model,

we may NEED to "shrink" the observed values. Deciding best how to do this is not easy (Tukey, 1979; Draper and Van Nostrand, 1979).

16. Efficiency compromises

One thing we must all be prepared to do, in and after the robust-resistant era, is to understand about how to compromise efficiencies—about what to do, at least qualitatively, when we want efficiencies in the face of quite diverse challenges. The natural leading case, of course, arises when (1) the y_i all estimate the same quantity and are uncorrelated, (2) against one challenge $\Sigma W_i y_i$ is the minimum variance estimator, where $\Sigma W_i = 1$ and (3) against a second challenge $\Sigma V_i y_i$ is the minimum variance estimator, where $\Sigma V_i = 1$. A natural measure of unpleasantness to minimize is

$$\lambda \frac{\text{variance for challenge 1}}{\text{minimum variance for challenge 1}} + \frac{\text{variance for challenge 2}}{\text{minimum variance for challenge 2}}$$

where λ may be 1.

A little algebra shows this is minimized, when

$$\text{weight on } y_i = \text{constant} \frac{V_i W_i}{\lambda V_i + W_i}$$

$$= \frac{\text{constant}}{\lambda/W_i + 1/V_i}$$

that is, when the reciprocal of the compromise weight is proportional to a weighted mean of reciprocals of the individual weights. Crudely, then, large compromise weights are only to be applied when the weights for both challenges—more generally, for all challenges—are large. (Matters are not always this simple. Weighting related only to value seems, like order statistics, to behave like a correlated case. If we go over to weighting gaps, however, as Guarino (1981) has shown by example, this principle appears to work well.)

The general principle suggested by this leading case is important throughout data analysis, and can be applied, at least qualitatively or semiquantitatively, in almost any data analytic situation.

17. General CDAPs

Critical data analysis ordinarily uses differences among pieces as a direct basis for assessing the variability that must be allowed for in looking at typical values and other summaries. (Geoffrey Jowett is working on more appropriate techniques for agriculture and husbandry data where the pieces are years.) The simplest case, "flatness", arises when the ADAP chosen ensures that the results for several pieces are related to those for single pieces in exactly the same way as means are related to individual values. Here we may use classic one-sample procedures—sign tests, Wilcoxon sums, Student's t or modern analogs (Gross, 1973, 1976, 1977; Kafadar, 1979). The effect of population shape on what population characteristic is most often enclosed by Student's t-intervals has been assessed for samples of 5–20 by Arthur (1979).

7 AN OVERVIEW OF TECHNIQUES OF DATA ANALYSIS

The next problem arises when we deal with "non-flatness" (cf. Beale, 1960; Efron, 1975), usually to be seen in the form of bias (for n pieces as compared to many) often conveniently thought of in the form

$$\text{ave}\{\text{result for } n \text{ pieces}\} = A + \frac{1}{n}B + \frac{1}{n^2}C + \cdots$$

Consider the calculation of pseudovalues—or pseudomeans—of the form

$$\frac{n\{\text{result for } n\} - m\{\text{result for } m\}}{n - m}$$

whose average is easily found to be

$$A + O \cdot B - \frac{1}{nm}C + \cdots$$

indicating one advantage—reduced higher-order bias—for $m = n - 1$, which corresponds to the jackknife (for reviews see Miller, 1974a, b).

Biases in variance are also relatively well-controlled in this way, probably also best by the jackknife. Since the detailed form of the jackknife was chosen to make averages and variances nearly correct, Student's t works more trustworthily with such pseudovalues than, for instance, rank-based procedures.

The weakness of jackknifing lies in "narrow" estimation, where only a few pieces combine to control the estimate. If this comes from a few giants among many pygmies, only stratification can help. If narrow estimation comes from the use of order statistics, simple functions of order statistics etc. the most reasonable solution seems to be the use of near-orthogonally intersecting sets of half-samples, for which non-flatness correction by pseudomeans is easy, though not usual.

Recently, Efron (1979) has advocated the use of bootstrap techniques where the n pieces are promoted to an infinite population of pieces ($1/n$th identical with each of the n observed) from which repeated samples of n are to be taken. (This is equivalent to sampling with replacement from the observed pieces.) The variability of a result among these repeated samples is then used as an estimate of the variability of that result for the original sample. This method appears to have no easy adjustment for non-flatness and fails when the desired result is sensitive to one piece being very similar to another. It does seem to have pedagogical advantages in communicating with the innocent.

As of this writing, the whole area merits considerably more study. Again as of this writing, best practice would seem to be to use the jackknife, weighted if appropriate,

(i) unless the results are order-statistic-like ("narrowly" estimated), when interpenetrating half-samples with pseudomeans should be used,

(ii) or unless there is a specific procedure for the particular result considered that is KNOWN to be robust/resistant.

Notice (1) that if the ADAP is flat, all these will degenerate into using Student's t on the results for individual pieces and (2) that the overall result will only be robust/resistant if the ADAP itself is robust/resistant.

Extension of all of these techniques to situations where the pieces are more highly structured as feasible. Such extensions should be used more often.

C. TOPICAL AREAS

The arrangement of this part, according to relatively conventional categories of statistical methods, is a confession of inadequately hard thinking about how data analysis would best be structured in 1980.

18. Re-expression

While we can make many of our data-analytic techniques robust and resistant enough to live calmly with rather messy deviations from conventional stochastic assumptions, such advances do little if anything to help us deal with deviations from traditional functional behavior. Fitting straight lines to unstraight relationships or additive decompositions to nonadditive two-way tables is little helped by resistant/robust techniques. Re-expression of responses (or also, in the regression case, of carriers), may, by cntrast, be very valuable in dealing with such difficulties.

Techniques for choosing such re-expression that combine robust-resistant smoothing with graphics-suggested arithmetic are now being actively developed; and show real promise (Tukey, unpublished).

Techniques for diagnosing needs for re-expression and suggesting appropriate forms have been growing in recent years. (For one account see Emerson and Stoto, 1981). We now have techniques for altering responses to promote additivity or straightness of plot or, in the multiple-regression situation, general quality of fit. Meeting such criteria related to functional behavior, when applicable, dominates meeting the other criteria, such as those for promoting constancy of spread or symmetry, that relate to statistical aspects of the model.

When important considerations—those of functional behavior—do not dictate how responses *et al.* should be expressed, we can also use re-expression to improve stochastic characteristics. This use, sometimes quite convenient and valuable, must however be thought of as secondary.

19. One and two batches

In classical Normal-theory methodology one summarizes a batch of data by two numbers—the mean and the standard deviation. Many texts suggest alternative pairs of statistics, for use when Normality cannot be assumed *a priori*. Much effort has gone in recent years to developing resistant estimates of location, and rather less on the corresponding scale problem.

However, in exploratory work a two-number summary is often inadequate. Batches are often highly skewed, and often contain one or more wild observations; it is important to be able to recognize these effects as they occur. Five numbers seems to be the minimum for an effective routine first-pass technique for summarizing one-dimensional batches. In EDA (Tukey, 1977), the five quantities are taken to be the median, two quartiles and two extremes, and a device called the box-plot is advocated as a routine graphical display.

But why stop at five? Some batches exhibit complex structure, that cannot be described adequately by these five quantities. EDA (Tukey, 1977) also suggests a number of elaborations of the basic idea in which remote outliers are listed individually. The challenge is to develop a technique that achieves flexibility while remaining simple, so that the resulting summaries are still easy to interpret. As yet we have no organized understanding of how to go about designing such a technique; this is still an art.

For comparing two or more batches, one can simply present and compare the respective five-number summaries. To calibrate the pairwise comparisons, a number of variations on the box-plot idea have been developed by Larsen *et al.* (1978).

When dealing with one or a few batches we not infrequently want to go beyond measures of location (centering) and scale (width). The classical measures based upon sample moments already fail for location and scale, to a varying degree, as we become more realistic about the heavy (stretched) tails of realistic distributions. The use of higher moments, presumably to reveal distribution shape, has almost nothing to recommend it. Assessment of g and h in

$$y = A + B \frac{\exp(gz) - 1}{g} \exp(hz^2/2)$$

which converts a unit Gaussian z into an observed y (with $h \geq 0$, or very nearly so) offers a mechanism to assess both a skewness and a elongatedness (Tukey, 1974a). The usefulness of this representation is more reasonable once we recognize that any distribution can be obtained as the distribution of

$$y = A + B \frac{\exp[g(z^2)z] - 1}{g(z^2)} \exp[h(z^2)z^2/2]$$

for functions $g(z^2)$ and $h(z^2)$ such that y is a monotone function of z.

Because (i) median $y = A$ and (ii) $dy/dz = B$ when $z = 0$, the values of g and of h are dimensionless quantities, and are invariant (against linear transformations).

20. Regression

Regression is the workhorse of statistics, the basic methodology for assessing the dependence of one variate on another. As has been shown by Daniel and Wood (1971), much can be achieved with a pre-packaged batch least-squares linear regression program, having a moderately large number of output options. However in recent years many new devices have been developed, each of which adds flexibility of approach and provides new kinds of insight.

As a result no one has a relatively comprehensive—but still manageable—package. Among devices for examining a single fitted regression and the corresponding residuals we have:

(i) Simple lists of fitted values and residuals, sorted by the independent variables, by fitted value (any of these may be helped by smoothing), and by value of residual (cf. Daniel and Wood, 1971).
(ii) Plots of residuals against fitted values, against independent variables (both those already in the fitted regression and others not), and probability plots of the residuals, also, probably most importantly, "partial residual" or "component plus residual" plots (Anscombe and Tukey, 1963; Larsen and McCleary, 1972; Wood, 1973).
(iii) Displays identifying "high-leverage" design points, either singly (Hoaglin and Welsch, 1978) or in groups (Andrews and Pregibon, 1978).
(iv) "Near neighbor" analysis (Daniel and Wood, 1971) that attempts to estimate a scale of residual variability as free as possible from assumptions regarding the correct form of the regression.
(v) A special-purpose display that belongs in this list is that form of Daniel's Half-Normal plot (Daniel, 1959), in which the regression coefficients for a well-chosen set of carriers are compared with one another. Gnanadesikan and Wilk have extended this idea in several directions (Gnanadesikan, 1977).

Many ways have been suggested for making regression computations more resistant to outliers. We have:

(i) Huber's original M-estimate proposals, exactly analogous to his estimates in the location case; a special case is L_1 regression (Huber, 1973).

7 AN OVERVIEW OF TECHNIQUES OF DATA ANALYSIS   141

(ii) Bickel's L-estimate methodology (Bickel, 1973).
(iii) Jaeckel's slope estimate based on minimizing a resistant measure of residual dispersion (Jaeckel, 1972).
(iv) Koenker's regression quantiles (Koenker, 1978).
(v) Several authors have proposed ways of reducing the influence of data at off-cloud (high-leverage) x-vectors (Hampel, Mallows, Andrews, cf. Hill, 1977). This is often referred to as "bounded influence regression".

Now we turn to a class of procedures in which several alternative regression fits are compared with one another.

(i) Mallows' C_p and variations (Mallows, 1973) and Spjøtvoll's (1977) suggestions.
(ii) Allen's PRESS (e.g. Allen, 1974).
(iii) Ridge regression, which was originally proposed (Hoerl and Kennard, 1970) as an estimation technique that improves on least squares; see Draper and van Nostrand (1979) and Smith and Campbell (1980) for recent reviews. In our opinion the only clear merit of the technique is that it provides a graphical summary of the impact of prior information (provided ridging is toward prior information, rather than toward a thoughtless zero).
(iv) Progressive Huberizing (Denby and Mallows, 1977), by which sensitivity to the effect of down-weighting large residuals can be assessed.

More complex forms of regression

It is sometimes said that "analysis of variance is just a form of regression". The common occurrence of anovas with several distinct error terms and the absence of both theory and technology for regressions involving two or more error terms makes the failure of this statement, at least with today's notion of regression, quite clear.

Path analysis, founded by Sewall Wright, commented upon by Tukey (1954) (see also Wright, 1960) and more recently pushed hard in sociology (e.g. four chapters by various authors in Blalock, 1971) makes use of structured multiple dependences among several variables. In this, it is similar to the estimation of simultaneous linear equation systems in economics, see Aigner and Goldberger (1977). See also, more generally, Duncan (1975), Heise (1975) and Jöreskog and Sörbom (1979).

Here we may also mention some techniques that produce piecewise constant fits—AID (Sonquist and others, 1964, 1973) or piecewise linear fits—the Binary Regression Tree approach of Breiman *et al.* (Friedman, 1979).

21. ANOVA

In our experience most analysis of variance is a matter of exploratory data analysis, rather than one of critical, confirmatory analysis.

In 1956 Pearson demonstrated several graphical displays of structured data. Daniel's (1959) half-normal plot was an early graphical, exploratory technique that Gnanadesikan and Wilk have extended in several directions (Gnanadesikan, 1977). The multiple-comparison methods of Scheffé and Tukey enable a severely limited amount of data-snooping to take place, while staying completely within the classical mold. Algorithmic and critical techniques are available for a wide variety of situations and a modest beginning has been made on diagnostic techniques.

In considering complex (factorial) anova, either in a classical spirit or as a start toward a robust/resistant version, a paper by Green and Tukey (1960) proposed techniques that still seem both reasonable and widely overlooked.

For a while it seemed that the main elements in providing a robust/resistant version of the analysis of variance were, basically,

(i) introducing a robust/resistant centering procedure to summarize each fiber—each set of values where all factors but one were constant—and
(ii) providing each line of the analysis of variance with two "mean squares" not one—a *large* one which includes the exotic behavior of any one or few summaries of the given kind, and a *small* one which is appropriate for estimating how much variability is carried up to higher lines.

More recently it has become clear that these, while important, are dominated by the need to deal with factors which have only a few (2, 3, 4, ...) versions or levels in a substantially more complex way.

(Procedures that identify individual outliers and try to use classical "Normal Theory" techniques otherwise have not yet been shown to be good enough.)

22. Smoothing

When data are sufficiently plentiful, and the linearity assumption is sufficiently unattractive, smoothing techniques come into play. Tukey (1977) contains some suggestions; see also Beaton and Tukey (1974) and Mallows (1980). Cleveland and Kleiner (1975) have published a detailed algorithm that worked very successfully in the second air-pollution study described in Section 5.

Wahba and her colleagues have advocated polynomial splines vigorously—in addition to their presumed usefulness in the regression context, they are claimed to be useful in estimating probability density functions and spectral densities (Wahba, 1975, 1979). Some would regard polynomials as intrinsically dangerous and prefer less conventional splines. Anderssen et al. (1974), Lenth (1977) and Huber (1979) have begun to investigate robustified splines.

There seems to have been little done on diagnostic or critical procedures.

23. Time series

Adequate coverage here would require much more space than is available for the entire survey. We confine ourselves to making one crucial distinction and noting a very few recent advances.

Besides the classical economist's separation of time series into trend, cycle and irregular—for which a satisfactory model is still lacking, and for which neither diagnosis nor criticism seems to have been seriously attacked—and all its modifications, we recognize three quite different sorts of (relatively-formal) time series analysis.

Spectrum analysis proper covers all the analyses that have grown, either by generalization or successive modifications from the leading case of one or more purely Gaussian signals. These techniques are appropriately used where repeated realizations will have quite different time histories (at least in detail) but somewhat similar estimated spectra, and where also there is enough data and/or enough distinctiveness of frequency behavior to make looking at spectra useful. (It takes 300 degrees of freedom to get one lone significant figure—for any s^2 or any spectrum density estimate!) For the elementary case, Blackman and Tukey (1959) still covers the "indirect" technique of analysis. (Notice that Tukey (1980) has pointed out a new kind of aliasing, which can arise in incautious use of the "direct" technique, as with the FFT.)

The second category of analysis arises in connection with processes or ensembles where alternate realizations would have quite similar time histories—perhaps differing only by background noise and/or measurement noise. Claerbout (1976) is a very useful and enlightening reference.

The third category arises when the combination of amount of data with distinctiveness of behavior is weak enough for us to do no more than ask: "By what simple mechanism could the data we see have come from a white-noise process (one where y_t is uncorrelated with $y_{t-1}, y_{t-2}, \ldots$)?" Here we have a spate of acronyms:—AR, MA, ARMA, ARIMA and the like. The classic reference is Box and Jenkins (1970, 1976). Like many other accounts, this writes of *Gaussian* white noise, and uses the Gaussian case to guide the choice of analysis. Surely not all examples to which these techniques might be applied are Gaussian—and we have looked at remarkably few examples to see whether or not particular instances are. We can hope, of course, that "Gaussian" has played a verbal, decorative role rather than essential one — that these techniques do, indeed, work well in practice.

Kleiner et al. (1979) have suggested ways of estimating spectral densities resistantly, and Denby and Martin (1979), Martin (1979) etc. have studied the problem of fitting ARIMA models in the presence of noise. Mallows (1980) has begun a theoretical study of non-linear smoothers. Cleveland et al. (1979) have developed a kit of tools to aid in seasonal adjustment, including several novel graphical displays that aid in assessing the magnitude and stability of the trend, seasonal and irregular components that are derived from the data.

For examples of sophisticated analysis of meaningful problems see, for instance, the work initiated by Lii et al. (1976) on the bispectra of turbulence (there are nine such), and the work of Brillinger et al. (1976) on the non-linear transfer functions of nerve impulses in living organisms.

There has been much recent work on optical data analysis, much of which can be traced from one or more of Rosenfeld (1976), Andrews and Hunt (1977), Leighty (1977), Casasent (1977, 1978) or Thompson (1978).

While complex demodulation (e.g. Bingham et al., 1967) was first emphasized as a computationally-cheap approach to single spectra, a use in which it was superseded by the FFT, it is now widely and diversely used, particularly in connection with higher-order spectra.

24. Multiplicity

Issues about conclusions related to multiple comparisons and multiple determinations are widespread and will be of increasing concern. They are all basically matters of critical, or of careful confirmatory data analysis. The natural first approximation is the spreading of error rate, in suitable fractions, over the relevant comparisons or determinations or whatever. This is usually referred to by the name of Bonferroni (e.g. Dayton and Schafer, 1973) and, in simple cases, corresponds to multiplying an individual P-value by the number of individual values or comparisons.

If we want to control a "-wise" error rate, namely the chance of *one-or-more-errors* in analyzing a group, family etc., we can tighten our confidence limits somewhat, as compared to the confidence limits for the same numerical error rate on a "per" basis. Whatever confidence limits we choose on a simultaneous basis, we may always ask what individual error rates we are using. The ratio

$$\frac{\text{individual error rates for ``per'' simultaneous}}{\text{individual error rates for ``-wise'' simultaneous}}$$

describes quite well the loss or gain from choosing one kind of simultaneous rate or another. For conventional multiple comparison situations, this ratio is almost always less than 2^2 — and usually less than 2 (Tukey, unpublished).

In regression — and in related questions — where something like "subset selection" takes place, we need to be able to deal with the bias involved in taking at face value the quality of what has been selected to be apparently best. (A "regression effect" applies to selecting the high apparent reduction in sum of squares, as well as to selecting a treatment or strain with the apparent highest apparent value.) Some possibilities have now been suggested (Tukey, 1979).

25. Counted and categorical data

We shall not attempt to be complete here, listing only the diversity of approaches to "contingency tables" and noting the rise of modified chi-squares.

At least five sorts of approach to contingency tables need to be mentioned. Goodman has written a sequence of long and careful papers, now collected into a book (Goodman, 1978). Much of the work on "log-linear models" is summarized in the book by Bishop et al. (1975). A different line has been followed by Grizzle et al. (1969), Koch and others (1972, 1977). The importance of focussing on cuts rather than on cells, when the factors have ordered categories, has been emphasized by Tukey (1977, Chapter 15, and unpublished extension to more-way tables). Benzecri's l'analyse des correspondances (Benzecri et al., 1970, 1973, see also Section 32) takes a complete "contingency table" as its leading case. Software developments (e.g. Baker and Nelder, 1978) have eased the computations.

For log-linear and some related models, Pregibon (1979) has developed diagnostic tools and a method of robust fitting.

"Denominatorless" chi-squares, with each individual-cell contribution being the square of

$$\sqrt{[2+4(\text{number observed})]} - \sqrt{[1+4(\text{number anticipated})]}$$

(the 2 should be replaced by 1 if none were observed), have been introduced (e.g. Mosteller and Tukey, 1977), and some of their properties studied (Tukey, unpublished). Such modifications offer gains in simplicity of use, closeness to nominal distributions, and a kind of robustness (corresponding to not believing very much in very small anticipated values).

26. Events in time: survival

The statistical analysis of events has been particularly actively studied in connection with clinical trials. Methods elaborated in seemingly quite different ways, as by Cox (1972), by Peto and Peto (1972), by Gehan (1965) and by Mantel and Haenszel (1959), are quite closely related, but far from equivalent (cf. for instance, Lininger et al., 1979, Tarone and Ware, 1977; Prentice and Marek, 1979; Mantel, 1981 and references therein). We understand a fair amount about this area, although we have much to learn. We have algorithms and critical techniques, but little in the way of diagnostics. An adequate summary of recent work here would take many pages. For an introduction see Breslow (1975).

A major need is to learn more about the consequences of making either the models or the assessments—or both—more flexible.

27. Non-linear fitting

The decades since the development of high-speed computation have seen a substantially increased interest in and use of non-linear parametric models for data. A completely general definition of this class of models would need to fall back on the criterion implied by its name: those models for which the standard approach of linear models is not adequate. More usefully, perhaps, one can cite two classes of models which cover most of the current work: predictive models and general objective criteria. The first class includes models that are non-linear in the obvious sense; namely, we try to predict a set of observations $y_1,\ldots,y_N$ by functions $f_1,\ldots,f_N$, which depend on the parameters $(\theta_1,\ldots,\theta_p)$ of the model in a non-linear way. We are then likely to minimize the (weighted) sum of squares of residuals. The second, more general class simply says that we will choose some criterion $C(y_1,\ldots,y_n; \theta_1,\ldots,\theta_p)$ which may or may not depend only on the residuals $y_i - f_i$, such that, say, small values of C often involves probabilistic assumptions, but this need not be the case.

The main activity has concerned the practical problem of fitting good values for the disposable parameters. Some work, not always too successful, has tried to go beyond this to statistical techniques for assessing the models. The fitting techniques, and algorithms implementing them, are largely based on two related ideas: approximating the general predictor $f_i(\theta)$ by a linear model and approximating the general objective $C(\theta)$ by a quadratic function.

Quadratic approximation of $C(\theta)$ leads to the most helpful general algorithms, based on successive approximations to parameters that minimize $C(\theta)$, using values of C and possibly first and second derivatives as well. Available algorithms are summarized in Chambers (1977, Chapter 6); for those wanting to understand recent numerical work Dennis and Moreé (1977) is recommended. Linear approximations lead in two directions: to methods for fitting non-linear least-squares models (Chambers, 1977, section 6g) and to fixed-point methods which treat classes of non-linear models as iterative linear models. Several approaches have been tried here: Wold (1966) and Nelder and Wedderburn (1972) provide some basic ideas.

Work on statistical aspects of non-linear models remains far from satisfactory. Both quadratic and linear approximations lead to estimates of the sample variances and covariances for the parameter estimates (Chambers, 1977, section 6e), but naive reliance on these can be misleading. More general understanding of the models needs such techniques as Monte Carlo simulation, if the model has a probabilistic basis, and jackknifing, cross-validation or other subsampling of the data (Chambers, 1973). Further research, particularly on incisive graphical displays of the models, is much needed.

28. Other multiresponse techniques

The book *Methods for Statistical Data Analysis of Multivariate Observations* by Gnanadesikan (1977) describes a large number of techniques, which we shall not list in detail here. We draw attention to three recent innovations.

Several suggestions have been made, since Galton (1861, 1863), for representing multivariate data graphically, so that comparisons among observational units can be made by comparing formalized displays. Some examples are Anderson's "glyphs", the standardized flag symbols used on meteorological charts, Andrews' function plots and Chernoff's faces. In all of these devices the choice of the mapping from the data to the various features of the graphical display is left to the user; this choice has a very major influence on the effectiveness of the display. Kleiner and Hartigan (1980) have developed a mode of display in which the data-matrix itself guides the choice of mapping. The variates are clustered, using a hierarchical algorithm. This produces a tree-topology for the variates. Then, each observational unit is represented by drawing a tree with this same topology with the lengths of the branches being functions of the values of the corresponding variates for that unit. Thus variates that are highly associated in the data-set are presented in adjacent branches, growing and shrinking together.

A second recent innovation is Wachter's technique for displaying the singular values of a data matrix that has both its dimensions large; or equivalently for examining the eigenvalues (principal components) of a large covariance matrix, or the canonical correlations between two numerous sets of variates. Wachter (1978) has established that under very general conditions the empirical distribution of such a set of quantities converges (as both dimensionalities increase indefinitely) to a non-random limit. This makes available a set of graphical techniques, analogous to those for comparing samples to populations, including probability plotting, that can be used to examine the set.

Generalizations of several familiar one-dimensional rank statistics have been provided by Friedman and Rafsky (1979). Questions of interpretation remain to be studied.

D. STRUCTURAL ESCALATION: MULTIVARIATE TECHNIQUES

Under this heading we treat first, briefly, those, often less familiar, methods that involve multidimensional guidance. In the next part we will run briefly through some of the more common types of data structures where escalation is easily available. (Rereading Section 8 at this point is likely to be helpful.)

29. Factor analysis

Given a two-way table of responses to individuals by questions or characteristics, there are a considerable variety of factor analytic techniques; some operating on correlations, some on covariances and some operating on the original data tables, some factoring questions and others factoring individuals. In almost every case it makes sense to distinguish two parts of what has been done:

(i) Finding a lower-dimensional subspace that comes close—in the naive, higher-dimensional space of the initial information—to the data vectors.
(ii) Choosing a coordinate system in this lower-dimensional space.

Agreement on the first part is much greater than agreement on the second: both usually involve linear combinations (without further diagnosis).

In working with factors it is usually important to distinguish

(i) factor loadings (which make up the factor pattern) from
(ii) factor scores.

The number we would use to estimate a factor (scores) need not be—and usually are not—the numbers expressing how much an initially given response comes from the factor in question (loadings).

Classical factor analysts still seem to find the rather classical "varimax" and "quartimax" procedures useful as well as "oblimin" and its modifications; some writers (e.g. Hunter, 1980) now argue for cluster-directed factoring procedures.

Classical methods are almost all confined to the autonomic step. The cluster users seem to have begun to use diagnostics in a rather preliminary way.

For recent reviews and monographs see Lawley and Maxwell (1971), Mulaik (1972), Harman (1976) and Kruskal (1978).

Three-way factor analysis

The last two decades have seen considerable emphasis on three-way factor analysis, particularly by Tucker (e.g. 1964, 1972, three-mode), Carroll and Chang (1970, CANDECOMP) and Harshman (1970, 1972, PARAFAC). The CANDECOMP-PARAFAC approach constrains the nature of the fit sufficiently to eliminate the need for a rotation step (Harshman and Berenbaum, 1980). Empirical evidence is slowly accumulating as to how very frequently the more-constrained approach is effective.

A diversion

(A recent paper (Steiger and Schönemann, 1978) returns to the presumed indeterminacy of factor scores. It seems likely that the heart of the argument reduces to:

(i) If one can estimate parameters by setting "observed" equal to "expected = average value given parameters", one must do this.
(ii) If this, because of non-linearity, gives multiple solutions, one must admit oneself stuck.
(iii) When, in regression on many correlated variables for instance, one finds regression coefficients rather indeterminate but regression estimates rather precise, this is a (mild) catastrophe.

We find the first two points inadequately cogent; the last merely an emphatic statement that good estimation need not require good coefficients.)

30. Multidimensional scaling (MDS)

In this type of procedure we give both linearity and extensibility to points not included in the data set at hand, striving to reproduce, in some familiar space, usually a Euclidean one,

(i) the distances among the given individuals (or characteristics etc.)— metric MDS—or
(ii) the ordering among these separations (which may come from distances or from more qualitative assessment of proximity—non-metric MDS.

Multidimensional scaling has, so far as we can see, so far been heavily confined to the first autonomic fitting step. Successive MDSings into Euclidean spaces of higher and higher dimension are routinely practiced, but this is diagnosis in only the weakest—and most confused—sense. We seem not to have seen useful diagnostic tools that isolate "stress" to either regions of the data or to individual points. Early attempts at point-related diagnostics are said to have been unsatisfactory. Comparison of individual fitted and observed distances is part of good practice.

We need to notice the existence of well-established non-metric methods which only need an ordering—and not a measuring—of the distances involved. For recent reviews of MDS see Kruskal and Wish (1978), Carroll and Kruskal (1978) and Carroll and Arabie (1980).

Multi-mode data

The extensions of MDS to multi-mode data are of increasing importance. For recent reviews see Carroll and Wish (1974) and Carroll and Arabie (1980).

-ices

For a quarter of a century, Louis Guttman has urged the importance of recognizing a variety of simple structures for the positioning of variables—structures for which at least one interesting leading case could be given—radices, circumplices and other -ices. (For an introduction and bibliography see Shye (1978), which includes Shepard's statement (Shepard, 1978) that "...the circumplex has, despite my initial skepticism, subsequently emerged in various guises out of my own data and analyses...") It would seem that the real problem is for the experimenter or data collector to be three things: (a) smart enough to include the right variables, questions or characters, (b) adept enough to measure well and (c) energetic enough to gather sufficient data. If these conditions are met, the topology of the implied structure has a good chance of being important.

31. Cluster analysis

There appears to be even more kinds of cluster analyses than there are kinds of MDS. Much energy seems to have been expended on the design of clustering techniques for which retrodiagnosis will indicate that all the separations made are, in some sense, demonstrable. It is far from clear that this "reality" is really important in the practical situations where cluster analysis is a useful springboard for other types of analysis. At the moment we see, at least most frequently, either (a) clustering techniques which produce a single—hopefully "correct"—clustering or (b) a hierarchical procedure which can be taken as offering many clustering procedures of different "strengths". Exploration of what to do next, as an organized procedure, is just beginning. For recent reviews, see Cormack (1971), Sneath and Sokal (1973), Hubert (1974), Hartigan (1975) and Blashfield and Aldenderfer (1978).

Graphical techniques to expand and display clustering results have recently been suggested by Cohen et al. (1977) and by Dunn and Landwehr (1980).

A notable exception to an absence of diagnostic techniques can be found in the recent work of Hogeweg (see Koek-Norman et al. (1979) for an example, and Hogeweg (1976a, 1976b) for some of the methodology). She uses an iteratively reweighted process involving an agglomerative dendogram, optimal splitting levels by a criterion, and reweighting of the characters in accordance with how well they help to discriminate the current set of clusters. The results of this method appear to approach the choices made by human taxonomists. (As yet there seems to have been no trials of setting aside the large weight characters, and starting over again with those that remain.)

This basically iterative clustering procedure is supplemented by an oligothetic (few-character) recognition guide and a projection on two-dimensions which appears to provide an example of "topology-plus".

For a discussion of local confirmatory procedures and related references see Sneath (1977a, b, 1979).

Relations to multidimensional scaling

Very recently there has been a surge of interest in the relations between—and the possibility of hybrid versions of—cluster analysis and multidimensional scaling (Kruskal, 1977; Carroll and Arabie, 1980, new taxonomy; Carroll and Pruzansky, 1981). As a result, work on diagnosing which appears more appropriate in a specific case has begun.

32. L'analyse des correspondances

Benzecri and collaborators (1970, 1973, 1976/77) have developed a combination of attitude, approach and technique directed toward two-way tables of non-negative numbers, f_{ij}. The leading case on which the calculations can be thought of as based is one in which

(i) the f_{ij} are observed counts,
(ii) the rows—and the columns—represent a complete inventory of a

meaningful domain, not artificially restricted, or perhaps a larger, widely distributed sample of such a domain and
(iii) the entities counted are homogeneous, and are measured in appropriately comparable units (number of seats rather than numbers of auditoria, when studying theaters, for instance).

Benzecri (1973, pp. 21ff) lays considerable stress on these last two requirements, and then goes on (pp. 26ff) to offer useful ways—often involving the replacement of one column by two or more columns—in which the techniques can be extended to more general responses, including those that can be negative as well as positive.

The basic expressions leading to the analysis are of the form

$$f_{ij} = f_{i+\cdot}f_{\cdot+j}\left\{1 + \left|\sum_{\alpha}\right|\lambda_\alpha^{1/2}F_\alpha(i)G_\alpha(j)\right\}$$

in which f_{ij} is analogous to a cell count, while $f_{i+\cdot}$ and $f_{\cdot+j}$ are marginal totals. The result of finding—or retaining—say the first three terms to provide *common* coordinates for *i*s and *j*s through

$$i \to [F_1(i), F_2(i), F_3(i)]$$
$$j \to [G_1(i), G_2(j), G_3(j)].$$

The emphasis is then on interpreting the coordinates thus provided—for example, the second coordinate, provided by $F_2(i)$ and $G_2(j)$—with the aid of our understanding about whichever of $\{i\}$ and $\{j\}$ we understand best.

Emphasis is placed (Benzecri et al., 1973, p. 3) on the need of "sciences de l'homme" for methods other than the classical ones and on "nos problèmes typologiques". It is argued (Benzecri, 1973, p. 48) that, more than 95 % of the time, interpretation should focus on the factors found ($\alpha = 1, \alpha = 2, \ldots$) in the formal analysis, with only occasional use of rotation. (The display of pp. 468–469 of Benzecri et al. (1973) suggests to us the usefulness of a 30° rotation, however.)

A sentence from Benzecri (1969, p. 36) is even more explicit: "Therefore, the help of a computer is needed to apply to the data previously collected a set of quasi-universal computations or rather transformations which give them such a shape that the man of the field may unarbitrarily read on the output what was indecipherable in the input."

Clearly Benzecri's hope is for *a standard analysis* which gives useful results with the minimum of human intervention—and with no intention of following-on with diagnosis and reanalysis. It is only with Benzecri's stress on homogeneity and exhaustivity that anything of this sort can seem possible. (We suggest that, for example, many of the difficulties raised in Benzecri's blood chemistry discussion (Benzecri et al., 1973, p. 22) are related to an unwillingness to take logarithms.)

We have devoted rather more space to "l'analyse des correspondances" than we might, because of the limited accounts of it in English. Benzecri (1969) is perhaps the only authentic such report. (The paper of Hill (1974) deserves careful attention, but focuses its analysis on the first coordinate.)

Interrelationships

Relationships of l'analyse des correspondances to a wide variety of other approaches have been called to our attention by our friends. The approaches mentioned include 1-degree-of-freedom-for-non-additivity, SVD, chi-square, the vacuum cleaner (Tukey, 1962), weighted Eckart–Young decompositions, "optimal scaling", canonical correlation and its generalizations, association models and a weighted CANDECOMP approach that can be closely related to Lazarsfeld's latent class analysis. An adequate discussion of these and other relationships, which may have to await much further work, would probably require a monograph.

33. Stage of development

If we look at the multidimensional methods of structural escalation as a whole,

(i) factor analysis,
(ii) cluster analysis,
(iii) multidimensional scaling and
(iv) l'analyse des correspondances,

and ask where they are in some evolutionary development, the analogous stage(s) for the more classical methods seems to be those of Francis Galton and Karl Pearson. They are often new tools which seem to open up new fields of inquiry, but tools whose answers are thought of in the raw—to be taken as they appear, without benefit of standard errors or tests of significance.

Multidimensional scaling, with its severe computational limitations on the number of points to be considered, is perhaps most Galtonian. L'analyse des correspondances, with its emphasis on exhaustivity and its computer programs willing to accept much larger bodies of data, is perhaps most Pearsonian, closest to K.P.'s ideal of samples large enough so one could be sure.

All, however, seem, until very recently, to have shown little published interest either in diagnosis as a route to improved analysis or in the assessment of sampling variation as a guide to understanding what we have learned—as opposed to what we have seen. It is believed, however, that there is much more use of diagnostics than would appear from published work. Some workers insist that repeated cycles of diagnosis and reanalysis are essential (R. A. Harshman, personal communication). For recent use of critical analysis in factor analysis, see Clarkson (1979). It is to be hoped that we will see continuing progress in both directions, though such progress can hardly come easily.

34. Coordinates, metric structure, topology-plus or classification

There are qualitatively different ways to express structure in a multi-dimensional way.

(i) We can seek out coordinates—numerically valued functions of the data points, which have to be somehow defined in terms of what information we have for the data points. Dangers to be avoided here include:

  (a) A failure to recognize that what is important about a coordinate is what entities have the same value (not which are different).
  (b) A tendency to stop with *linear* combinations of *naively* chosen expressions of initially given information.
  (c) A failure to be clear just what changes of coordinates remain to be freely made: Nothing? Rigid rotations? All non-singular linear transformations? (Yes answers pertain to three quite different uses of coordinates!)

(ii) We can seek out a metric structure—in which the data points are locked together, but are freely subject to rigid motions. Dangers to be avoided here include:

  (a) Incomplete thought as to whether the metric structure is a step on the way to a possibly more useful structuring, or is an end in itself. (For example, if our points lie in a plane close to a circle, should we or should we not introduce polar coordinates?)
  (b) Failure to develop techniques for taking a metric structure and representing it—perhaps approximately—in terms of coordinates defined in terms of our original information.

(iii) We can seek out a topology-plus—going somewhat beyond the minimum that "topology" connotes, placing the data-points in a way that reveals more than just nearness or farness, providing some information about relative nearness of entities that are near. (This might correspond to an approximate *local* metric structure or to *locally* approximately rigid embeddings in a plane or other well-understood space.) The main danger to be avoided here is the awesome absence of techniques to be tried out and compared (but cf. Shepard and Carroll, 1966).

(iv) We can seek out a classification—in which lesser entities are grouped into greater entities, perhaps once, perhaps in repeated steps of hierarchy. Dangers to be avoided here include:

(a) An unwarranted belief that a particular situation can *surely* be well classified, so that there *must* be a technique that will help us.
(b) A belief that classification, particularly one step of an aggregation into (often small) groups can *only* help us if the boundaries of the groups are clear, sharp and well-defined.
(c) Forgetfulness that, when we have only a useful grouping, without sharp boundaries, we should not automatically attribute "reality" to the boundaries.

No approach is without its dangers. All these dangers should be taken seriously.

The Guttman effect

In both multidimensional scaling and l'analyse des correspondances" it is a familiar possibility that later "factors" will come close to being functions (nearly quadratic or cubic, sometimes) of earlier ones. Because of Louis Guttman's early emphasis on this possibility, Benzecri *et al.* (1970, 1973) refer to it as the Guttman effect. It clearly offers us a chance to make an interesting—and probably helpful—*nonlinear* change of coordinates.

An unrelated, but very suggestive analog arises in dealing with the simpler data of particle physics (Friedman *et al.*, 1980) where points are concentrated *near* intermediate dimensional manifolds, and our main task is to recognize and describe these manifolds, and then to clear away the associated points to see what is left.

Once we have fitted ourselves out with techniques for finding expressions of useful coordinates in terms of initial information, it is hard to see why we should be willing to be constrained by initial structure—whether from MDS, l'analyse des correspondances or anyhow. The natural step, it would clearly seem, is to *choose coordinates to simplify the description* of the phenomena.

The message from display

Efforts to study the most effective ways to visualize the appearance—or structure—of point clouds in three or more dimensions (Tukey and Tukey,

1981) in terms of what can be shown on a page or a screen, have led to simple, if uncomfortable, conclusions, namely:

(i) If we are prepared to look at enough different views (or make a solid model that we can handle and move, or use a dynamic computer-drive display) we can do quite well in appreciating point-cloud patterns in three dimensions.
(ii) To go further, it would seem that about the best we can do is to show the smooth, summarized part of the behavior in a few coordinates beyond the horizontal and vertical, in which we can, of course, show either full detail or summarization by agglomerated points.

The conclusion for structuring for display is clear:

(i) We should pack as much into a plane as we can.
(ii) After we have done this, we can do what we can with additional information, using our packed plane as a map.

The best way to pack as much in a plane as we can would seem to be *topology-plus*. Thus the message from display raises the urgency for finding good methods in this direction.

E. STRUCTURAL ESCALATION: SOME DATA STRUCTURES

As promised above, in this part we shall run rather rapidly through a variety of data-structures, mentioning the kinds of structural escalation available for each.

35. Batch of responses

Given only a set of y_i, not otherwise structured, only one form of structural escalation seems natural—separation by value into potentially distinct subgroups, whether (a) more or less distinctly distributed or (b) more or less outlying or inlying. Techniques routinely available here are not necessarily well-considered from the view taken in this part, since much too much effort has been spent on autonomic fits—clustering procedures, outlier identifiers—that go only far enough so that retrodiagnosis will sanctify all that we have done. Future methods will, we hope, break up the batch automatically into what are likely to be too many parts, and then do some gluing back together again.

36. Batch of (x, y) pairs

We have discussed above various aspects of fitting a straight line, as one of the classical examples. To what we have said, we need to add a variety of approaches to re-expression, including walking up and down the ladder of powers (e.g. cf. Tukey, 1977, chapter 6) and smelting, a form of smoothing by excision (Tukey, unpublished).

37. Row-by-column table of responses

Two-way analyses

We have mentioned some aspects here in connection with the second classical example. To these we have to add the techniques related or analogous to ODOFFNA (one-degree-of-freedom-for-non-additity) (Tukey, 1949), which focussed on retrodiagnosis in a narrowly enhanced fit, to the diagnostic plot (cf. Tukey, 1977, Chapters 10 and 11) which is more in tune with today's thinking (a) by being graphical, (b) by being less rigidly tied to a specific form of enhancement and (c) by emphasizing diagnosis rather than retrodiagnosis.

To this single-additional constant enhancement we have to add a whole array of enhancements involving fitting rather more constants (cf. Tukey, 1962; McNeil and Tukey, 1975).

For at least the simpler analyses, good critical techniques (not just F-tests) are classically available.

Factor analysis

If the structure of our table is at all like individual-by-character, one can try various versions of factor analysis. Diagnosis is not likely to be available. Criticism may be available to the extent of comparing the analysis of tables that share one margin. For a careful discussion see Tucker (1964, 1972).

l'Analyse des correspondances

If we are willing to use Benzecri's leading case, which means making or having non-negative entries which meet or simulate the additional conditions discussed in Section 29, we may take our row-by-column table of responses to l'analyse des correspondances. Neither diagnosis nor criticism seems to be available for the result.

Promotion to proximities

If we are willing to fix on "corresponding scales" for different responses or characters in an individual-by-response-or-character table we can generate a metric relating individuals. If we are willing to do this, or to fix a metric in some other way, we will have promoted our table to an (individual)² table of (metric) proximities, thus opening the way to MDS, cluster analysis etc.

Canonical analysis

If our table is, or resembles, an individual-by-character table and the individuals are classified into groups or populations, we can use the sensibly-displayed version of multiple-discriminant analysis commonly called canonical analysis (e.g. Ashton *et al.*, 1957; Delaney and Healy, 1964). (In doing this, we should be careful not to accept the coordinates as they come unless there is a critical analysis of differences among eigenvalues. Such a canonical analysis *is* a critical analysis, but there seems to be nothing available in the nature of diagnosis.

Circumstances-by-circumstances

When the two factors enter symmetrically with a single response—for instance, yields of a chemical process as a function of temperature and pressure, we are surely open to two-way analysis (perhaps to be followed by functional fitting if the factors are quantitative), which has often proved helpful. We could try promotion to proximities or l'analyse des correspondances—so far, experience has not made either something that seems natural to try routinely.

Quadrant II data

The case where the response is 0 or 1 includes the kinds of psychological data classified by Coombs (1964) as Quadrant II. In Quadrant IIa the 1 refers to "passes" or "exceeds"; in Quadrant IIb to "is acceptable" and the like. Coombs offers algorithmic data analyses for these cases, without diagnosis or criticism.

38. (Individuals)² or (circumstances)², proximities

Given a symmetrical table for which, either indirectly by promotion from a two-different-factor table or directly as in some psychological judgments, we have measures of *proximity*, we can go ahead with either cluster analysis or MDS. (Presumably also with topology-plus, were techniques available.) Again both diagnosis and criticism are largely absent.

39. Individual-by-(stimuli)^2

Two of Coombs's (1964) quadrants fall here. If the response to a pair of stimuli is a preference, we are in Quadrant I (Ia for forced preferences, IIb for do you prefer either). If the response is a "more than" vs. "less than" judgment, we are in quadrant III (IIIa for forced choice, IIIb for do they match). Algorithmic analyses are offered for all these, without diagnosis or criticism.

40. Three-way data structures in general

Particular attention seems to be being given recently to individuals-by-variables-by-occasions arrays and to individuals-by-stimuli-by-rating scales arrays. For general techniques see Sections 29 and 30. Diagnostic use of error-variance breakdowns are beginning to appear (e.g. Harshman *et al.*, 1977).

41. Individual × [(stimuli)^2]^2

The last of Coombs's quadrants falls here, Quadrant IV for individual judgments of differences in similarity (IVa for forced choice and IVb for do match.) Again algorithmic techniques are offered, without diagnosis or criticism.

F. CONCLUDING ISSUES

This shorter part deals with a set of issues that seem important for the near and middle future.

42. The future of data analysis

One of us wrote an article under this title almost two decades ago (Tukey, 1962). We shall indicate how its considerations relate to the present-day situation. A few recurrent comments are conveniently abbreviated as follows:

(a) = General emphasis of comments holds today as they held then.
(b) = Overtaken by M- and W-estimates.
(c) = Not developed enough—see particularly end of Section 21 (comments on rob/res anova).
(d) = Growth of tools seems to have outstripped growth of use.

Going through in order (section numbers here refer to Tukey (1962)), we find: (a) for I (General considerations). (b) for II (Spotty data). In part III, (a) for general emphasis of sections 16 and 17; (b) for section 18; (c) for sections 19 and 20. (a) for IV (Multiple-response data), but see also Parts D and E above on structural escalation. In part V, spectrum analysis has greatly proliferated, but the other issues of section 27 have made little progress; selection and screening (section 28) need a review we are not supplying here; (d) for sections 29 and 30; (b) for section 31; the Behrens–Fisher (and Smith–Welch–Fisher) problem (section 32) deserves a review we are not giving. (a) for VI (Flexibility of attack) with some progress re section 33; indication and conclusion more or less transmuted into EDA and CDA; and a need for expansion and propagation for sections 36 and 37. (c) for VII (The vacuum cleaner), also note relationship to (not identity with) l'analyse des correspondances. (a) for VIII (How shall we proceed?), although section 47 undervalued substantially what the computer—maxi, mini or hand-held—has since become and will become in the near future.

43. General state of the field

The analysis of data, in some form, is probably older than the Egyptian pyramids, since early prediction of eclipses, e.g. by "saros" cycles, must have been based upon numerical study of observations. Today, many techniques are recognized only in specific fields, while others are widely used, perhaps without identification as such. A picture of points and a "fitted" straight line is perhaps the most widespread technique, while Student's t, used for significance or confidence, also a widespread example, is of quite a different kind.

We have not yet gone about structuring the field as a whole in an understandable and effective way. As we do, we will find ourselves paying attention to less-familiar facets, such as the distinction among:

(1) Techniques used to show that we *don't* need to go further (in our process of partial description).
(2) Techniques used to show that we *do* need to go further.
(3) Techniques used to show that we do need to go further, and in about *what* direction.

(4) Techniques used to show the stability or instability of the description we have reached.

(In a real sense, data analysis is the art of finding illuminating descriptions that are both partial and quantitative.)

We have large tasks before us, both in developing initial structure and in using this structure to organize what others have done and to see what still others might do.

44. Half-formalizing an art

Up to the present, data analysis, as a practical procedure, has been an art (or craft)—more precisely a bundle of arts and crafts, some rather general, others confined to quite specific areas of application. Those of us who recognize the importance of more effective data analysis must feel the urgency of transforming it somewhat more nearly into an organized body of knowledge. (Some seem to think that it could be fully formalized and/or mathematized within a few years; others would doubt the value of thinking about such complete formalization, even as an unattainable goal.) Doing this even halfway, which might be a goal acceptable to all, will not be at all easy, and will require our best skills in dealing with incompletely specified and inchoate problems.

Yet we need to do just this if we are to mobilize the powerful forces of graduate lectures, professorial research and Ph.D. theses for the accelerated development of both individual techniques and overall structures.

We need to be able to structure data analysis in a more effective way—in a way where we can preserve both flexibility and a backbone. We need recognizable styles of study; and standards against which studies can be judged for publication. No one of these things will be easy.

45. Surveying and fine-tuning

One type of activity important in developing data-analysis as an organized body of knowledge is surveying the varied fields of application to learn what methods are in fact used for similar purposes, perhaps rather similar, perhaps nearly identical. A necessary part of such an inquiry is careful consideration of the extent to which purposes are indeed the same, and careful discrimination between alternative purposes when they are different. We will need to develop standards against which such inquiries can be usefully evaluated.

An important corollary to such an inquiry, possibly as a part of the same study, but possibly not, is the "fine tuning" of the more desirable methods to make them still better. We need well thought-through guidelines to help us decide to what extent fine-tuning should be on the basis of pure simulation, to what extent on the basis of real data used as is, and to what extent on the basis of simulations constructed using real data.

46. How adequate will the computer be, how soon?

We can never expect that the computer will do everything, but we can always plan for it to do more. The ultimate choice can be either of two extremes: (a) a single large block of computation with massive output to be whipped into shape by extended human effort, or (b) human interaction at every possible step. We have to expect that the computation will be in moderate-sized blocks, blocks that involve *internal* selection of which things are to be output of the many that are computed, blocks that are separated by human examination, human judgment and human choices.

This means using algorithms both

(i) to produce alternative preoutputs (potential outputs)—autonomic, diagnostic or critical—and
(ii) to select which among these already produced preoutputs should in fact be shown to the analyst.

There has been much effort on autonomic algorithms (but rarely as explicit alternatives), some effort on diagnostic algorithms, rather more on critical algorithms and very little on selective algorithms.

Indeed, little if any thought seems to have been given to truly selective processes of data analysis—SDAP. The nearest we seem to have come seems to be:

(i) IDAP—iteratively reweighted DAP, which includes Pauline Hogeweg's clustering processes as well as most modern robust/resistant techniques.
(ii) LDAP—algorithms for the k-best (a), for each (b), as in Furnival and Wilson's (1974) algorithms for the k-closest subset regressions for each size of subset or Friedman *et al.* (1977) algorithm for the k-nearest neighbors of each point in a multidimensional cloud.

Still further from the selectivity are

(iii) HDAP—hill-climbing processes such as projection pursuit (Friedman and Tukey, 1974).

These are not truly selective, since they select during computation rather than after.

7 AN OVERVIEW OF TECHNIQUES OF DATA ANALYSIS 163

Clearly we need to start working much harder on both diagnostic and selective algorithms. In doing this we need to avoid the idea that all we need to do is to pick a rather arbitrary criterion and then work on implementing this criterion in an algorithm. The hard work has to come in choosing a criterion—or perhaps in choosing an algorithm whose criterion would be hard to write down. In doing this we would often want to look carefully at 100 or more results of a group of ADAPs to see which should be diagnosed or selected. Then we could compare criteria already thought of and learn how to invent new criteria in terms of their performance on this interpreted set. (It is likely that it will be important to have the set of results making up the test bed interpreted by a number of data analysts of different attitudes and predictions.) Only after we have done some of this can we hope to estimate how adequate the computer is likely to become in the foreseeable future.

47. Styles of report

Why is it that arguments based on statistical analysis so often fail to convince? Three reasons can be advanced. First, the procedures used may not have been explained adequately. Confidence intervals are often misinterpreted as being statements about posterior probability. (Clearing this point up may not make the results more palatable!) Second, critical procedures from the classical repertoire (both frequentist and Bayesian) are often applied in situations where the conventional assumptions have not been checked, so that the apparent precision of the inferences at least appear to be in danger of being fallacious. Perceptive non-professionals feel uncomfortable (often correctly) with the idea that you can get something for nothing, while the data analysts have not always explained adequately how much or how little attention each of the conventional assumptions deserves. Third, comes the pervasive difficulty that the relevance of the numerical data that is available to the substantive issue under discussion is often not clear (even if the substantive issue itself is clear!).

Perhaps little can yet be done about this third difficulty, but the data-analytic approach does have something to contribute with respect to the first two. Many good techniques are simple and direct, with transparent relevance, for example the box-plot and the scatter-plot. Many of the procedures enable direct comparisons to be made between pairs of data-sets or among parts of a single data-set. Randomness is often a mysterious assumption that by some unexplained magic enables precise statements of probability to be made, but variability is a simple observable fact that can often be displayed and understood directly.

These considerations have bearing on the issue of statistical reporting. At one level, that of the "executive summary", where only the conclusions are needed, there is no need to explain the details of the analysis, or even, sometimes the nature and source of the data. However, two other kinds of report are needed. One is the report for the subject-matter expert, who is interested in understanding in detail the conclusions reached, the data they are based on and the reasoning involved. Here the statistician has a deep obligation to explain exactly what it is that he did, so that the client can follow the steps of the analysis in detail. In saying "what", however, much will have to be done by reference, just as a chemist will refer elsewhere for the description of the analytical method. Such references will rightly excuse the discussion of arithmetic details, but they will enhance the responsibility for spelling out the true purposes of the steps and of whatever we know about their behavior/misbehavior that is likely to be relevant.

A major difficulty here is explaining how the data was explored. It is hard to map a flexibly followed exploration and harder to describe the map tersely and effectively. We need to be prepared, whether we are analysts or subject-matter experts (or both) for the data to be broken up into many pieces of possible interest and then to be partly re-assembled into the decomposition whose description and evaluation is our final analysis. When we evaluate critically, when we move toward significance or confidence rather than pure indication, we will almost always have to face problems of multiplicity that challenge our present understanding. We need to be as honest and explicit about multiplicity as we should be about assumptions—trying desperately to thread our way between the Scylla of overoptimism and the Charybdis of overcaution. Such a report cannot be written without the statistician becoming at least somewhat familiar with the subject-matter.

The final type of report is that written for data-analytic and the statistical professions. Here it is difficult to say what should be done, since so few examples of good analysis have been published. As yet we do not have adequate language in which to communicate nuances of data-analytic technique, without reference to the subject-matter. Perhaps this will come.

48. Acknowledgments

We thank J. M. Chambers for a draft of Section 27. We are grateful to D. F. Andrews, D. R. Brillinger, J. D. Carroll, D. R. Cox, C. Daniel, W. J. Dixon, J. L. Dolby, J. H. Friedman, I. J. Good, J. C. Gower, R. A. Harshman, J. B. Kruskal, W. H. Kruskal, D. Pregibon, N. Reid, H. V. Roberts, J. T. Smith, S. M. Stigler, G. S. Watson and W. T. Williams who made comments on an earlier version of this paper. These were very helpful to us, but our errors, omissions and infelicities remain our own responsibility.

References

Aigner, D. J. and Goldberger, A. S. (1977). *Latent Variables in Socioeconomic Models.* North-Holland, New York.

Allen, D. M. (1974). The relationship between variable selection and data augmentation and a method for prediction, *Technometrics,* **16,** 125–127.

Anderssen, R. S., Bloomfield, P. and McNeil, D. R. (1974). Spline Functions in Data Analysis. Technical Report 69 (Series 2), Department of Statistics, Princeton University.

Andrews, D. F. (1978). Data analysis, exploratory. In *International Encyclopedia of Statistics* (W. H. Kruskal and J. M. Tanur, eds), 97–106. The Free Press, New York.

Andrews, D. F. and Pregibon, D. (1978). Finding the outliers that matter, *Journal of the Royal Statistical Society* B, **40,** 85–93.

Andrews, D. F., Bickel, P. J., Hampel, F. R., Huber, P. J., Rogers, W. H. and Tukey, J. W. (1972). *Robust Estimates of Location.* Princeton University Press, Princeton.

Andrews, H. C. and Hunt, B. R. (1977). *Digital Image Restoration.* Prentice-Hall, Englewood Cliffs.

Anscombe, F. J. and Tukey, J. W. (1963). The examination and analysis of residuals, *Technometrics,* **5,** 141–160.

Arthur, S. P. (1979). Skew/stretched Distributions and the t-statistic. Ph.D. thesis, Princeton University.

Ashton, E. H., Healy, M. J. R. and Lipton, S. (1957). The descriptive use of discriminant functions in physical anthropology, *Proceedings of the Royal Society* B, **146,** 552–572.

Baker, R. J. and Nelder, J. A. (1978). *The GLIM System, Release 3.* Numerical Algorithms Group, Royal Statistical Society.

Beale, E. M. L. (1960). Confidence regions in non-linear estimation, *Journal of the Royal Statistical Society* B, **22,** 41–76 (discussion, 76–88).

Beaton, A. E. and Tukey, J. W. (1974). The fitting of power series, meaning polynomials, illustrated on band-spectroscopic data, *Technometrics,* **16,** 147–185 (discussion, 187–192).

Benzecri, J. P. (1969). Statistical analysis as a test to make patterns emerge from clouds. In *Methodologies of Pattern Recognition* (S. Watanabe, ed), 35–74. Academic Press, New York.

Benzecri, J. P. (1976/77). Histoire et préhistoire de l'analyse de données, *Les Cahiers de l'Analyse des Données* (Partie I, La Préhistoire) **1,** 9–32; (Partie II, La Biometrie) **1,** 101–120; (Partie III, Era piscatoria) **1,** 221–241; (Partie IV, La Psychometrie) **1,** 343–366; (Partie V, L'analyse des correspondances) **2,** 9–40.

Benzecri, J. P. *et al.* (1970). *L'Analyse des Données, I. La Taxonomie.* Dunod, Paris (revised 1976).

Benzecri, J. P. *et al.* (1973). *L'Analyse des Données, II. L'Analyse des Correspondances.* Dunod, Paris.

Bertin, J. (1973). *Semiologie Graphique,* 2nd edn. Mouton, Paris.

Bertin, J. (1977). *La Graphique et le Traitement Graphique de l'Information.* Flammarion, Paris.

Bertin, J. (1980). *Graphics and the Graphical Analysis of Data.* DeGruyter, Berlin. (Translated by W. Berg of J. Bertin, 1977).

Bickel, P. J. (1973). On some analogues to linear combinations of order statistics in the linear model, *Annals of Statistics,* **1,** 597–616.

Bingham, C., Godfrey, M. D. and Tukey, J. W. (1967). Modern techniques of power spectrum estimation, *Trans. IEEE Audio Electroacoustics*, **AU-15**, 56–66.
Bishop, Y. M. M., Fienberg, S. E. and Holland, P. W. (1975). *Discrete Multivariate Analysis: Theory and Practice*. MIT Press, Cambridge, Massachusetts.
Blackman, R. B. and Tukey, J. W. (1959). *The Measurement of Power Spectra*. Dover, New York.
Blalock, H. M. (ed.) (1971) *Causal Models in the Social Sciences*. Aldine-Atherton, Chicago.
Blashfield, R. K. and Aldenderfer, M. S. (1978). The literature on cluster analysis, *Multivariable Behavioral Research*, **13**, 271–295.
Box, G. E. P. and Jenkins, G. M. (1970) (1976, 2nd edn.). *Time Series Analysis: Forecasting and Control*. Holden-Day, San Francisco.
Breslow, N. E. (1975). Analysis of survival data under the proportional hazards model, *International Statistical Reviews*, **43**, 45–58.
Brillinger, D. R., Bryant, H. L. Jr. and Segundo, J. P. (1976). Identification of synaptic interactions, *Biology of Cybernetics*, **22**, 228–238.
Campbell, N. R. (1919). *Physics: The Elements*. Cambridge University Press, Cambridge.
Campbell, N. R. (1957). *Foundations of Science*. Dover, New York (reprint of Campbell, 1919).
Carroll, J. D. and Arabie, P. (1980). Multidimensional scaling, *Annual Review of Psychology*, 607–649.
Carroll, J. D. and Chang, J. J. (1970). Analysis of individual differences in multidimensional scaling via an N-way generalization of Eckart-Young decomposition, *Psychometrika*, **35**, 283–319.
Carroll, J. D. and Kruskal, J. B. (1978). Scaling, multidimensional. In *International Encyclopedia of Statistics* (W. H. Kruskal and J. M. Tanur, eds), 892–907. Free Press, New York.
Carroll, J. D. and Pruzansky, S. (1981). Discrete and hybrid scaling models. In *Similarity and Choice* (E. D. Lanterman and H. Feger, eds). Hans Huber, Bern.
Carroll, J. D. and Wish, M. (1974). Models and methods for three-way multidimensional scaling. In *Contemporary Developments in Mathematical Psychology* (D. H. Krantz, R. C. Atkinson, R. D. Luce and P. Suppes, eds), Vol. 2, 57–105. Freeman, San Francisco.
Casasent, D. (ed.) (1977). *Optical Signal and Image Processing*, Proceedings of the International Optical Computing Conference of '77. (SPIE volume 118) IEEE, New York.
Casasent, D. (ed.) (1978). *Optical Data Processing: Applications*. Springer, Berlin.
Chambers, J. M. (1973). Fitting nonlinear models: numerical techniques, *Biometrika*, **60**, 1–13.
Chambers, J. M. (1977). *Computational Methods for Data Analysis*. Wiley, New York.
Chambers, J. M., Cleveland, W. S., Kleiner, B. and Tukey, P. A. (1982). *Graphical Methods for Data Analysis* (in preparation).
Claerbout, J. (1976). *Fundamentals of Geophysical Data Processing*. McGraw-Hill, New York.
Clarkson, D. B. (1979). Estimating the standard errors of rotated factor loadings by Jackknifing, *Psychometrika*, **64**, 297–313.
Cleveland, W. S. and Kleiner, B. (1975). A graphical technique for enhancing scatterplots with moving statistics, *Technometrics*, **17**, 447–454.
Cleveland, W. S., Dunn, D. M. and Terpenning, I. J. (1978). SABL: A resistant

seasonal adjustment procedure with graphical methods for interpretation and diagnosis. *Proceedings of the Conference on Seasonal Analysis of Economic Time Series*, 201–241. Economic Research Report ER-1, US Department of Commerce: Bureau of the Census.

Cohen, A., Gnanadesikan, R., Kettenring, J. R. and Landwehr, J. M. (1977). Methodological developments in some applications of clustering. In *Applications of Statistics* (P. R. Krishnaiah, ed.), 141–162. North-Holland, New York.

Coombs, C. H. (1964). *A Theory of Data*. Wiley, New York.

Cormack, R. M. (1971). A review of classification, *Journal of the Royal Statistical Society* A, **134**, 321–353 (discussion 353–367).

Cox, D. R. (1972). Regression models and life-tables, *Journal of the Royal Statistical Society* B, **34**, 187–202 (discussion, 202–220).

Cox, D. R. and Snell, E. J. (1968). A general definition of residuals, *Journal of the Royal Statistical Society* B, **30**, 248–265 (discussion, 265–275).

Daniel, C. (1959). The use of half-normal plots in interpreting factorial two-level experiments, *Technometrics*, **1**, 311–341.

Daniel, C. and Wood, F. S. (1971). *Fitting Equations to Data*. Wiley, New York.

Dayton, C. M. and Schafer, W. D. (1973). Extended tables of t and chi-square for Bonferroni tests with unequal error allocation, *Journal of the American Statistical Association*, **68**, 78–83.

Delaney, M. J. and Healy, M. J. R. (1964). Variation in the long-tailed field mouse (*Apodemus sylvaticus* (L.)) in northwest Scotland. II. Simultaneous examination of all characters, *Proceedings of the Royal Society* B, **161**, 200–207.

Denby, L. and Mallows, C. L. (1977). Two diagnostic displays for robust regression analysis, *Technometrics*, **19**, 1–13.

Denby, L. and Martin, R. D. (1979). Robust estimation of the first-order autoregressive parameter, *Journal of the American Statistical Association*, **74**, 140–146.

Dennis, J. E., Jr. and Morée, J. (1977). Quasi-Newton methods, motivation and theory, *SIAM Review*, **19**, 46–89.

Draper, N. R. and Van Nostrand, C. (1979). Ridge regression and James–Stein estimation: review and comments, *Technometrics*, **21**, 451–466.

Duncan, O. D. (1975). *Introduction to Structural Equation Models*. Academic Press, New York.

Dunn, D. M. and Landwehr, J. M. (1980). Analyzing clustering effects across time, *Journal of the American Statistical Association*, **75**, 8–15.

Efron, B. (1975). Defining the curvature of a statistical problem (with applications to second-order efficiency), *Annals of Statistics*, **3**, 1189–1242.

Efron, B. (1979). Bootstrap methods: another look at the jackknife, *Annals of Statistics* **7**, 1–26.

Emerson, J. D. and Stoto, M. A. (1981). Exploratory methods for choosing power transformations. To appear in *The Journal of the American Statistical Association*.

Fisher, R. A. (1926). The arrangement of field experiments. *Journal of the Ministry of Agriculture*, **33**, 503–513. (Reprinted as paper 17 in Fisher, R. A. (1950). *Contributions to Mathematical Statistics*. Wiley, New York.)

Friedman, J. H. (1979). A tree-structured approach to nonparametric multiple regression. In *Smoothing Techniques for Curve Estimation*. (Th. Gasser and M. Rosenblatt, eds), 5–22. Springer, Berlin.

Friedman, J. H. and Rafsky, L. C. (1979). Multivariate generalizations of the Wald–Wolfowitz and Smirnov two-sample tests, *Annals of Statistics*, **7**, 697–711.

Friedman, J. H. and Tukey, J. W. (1974). A projection pursuit algorithm for exploratory data analysis, *IEEE Transactions on Computations*, **C-23**, 881–890.

Friedman, J. H., Bentley, J. L. and Finkel, R. A. (1977). An algorithm for finding best matches in logarithmic expected time, *ACM Transactions on Mathematical Software*, **3**, 209–226.

Friedman, J. H., Tukey, J. W. and Tukey, P. A. (1980). An approach to data analysis for data that concentrate near higher-dimensional manifolds. In *Data Analysis and Informatics* (E. Diday et al., eds). North-Holland, Amsterdam and New York.

Furnival, G. M. and Wilson, R. W. (1974). Regressions by leaps and bounds, *Technometrics*, **16**, 499–511.

Galton, F. (1861). Meteorological Charts, *Philosophical Magazine* (4th ser.), **22**, 34–35.

Galton, F. (1863). *Meteorographica or Methods of Mapping the Weather; Illustrated by Upwards of 600 Printed and Lithographed Diagrams Referring to the Weather of a Large Part of Europe During the Month of December 1861*. Macmillan, London.

Gehan, E. A. (1965). A generalized Wilcoxon test for comparing arbitrarily singly censored samples, *Biometrika*, **52**, 203–223.

Gnanadesikan, R. (1977). *Methods for Statistical Data Analysis of Multivariate Observations*. Wiley, New York.

Goodman, L. A. (1978). *Analyzing Qualitative/Categorical Data: Log-linear Models and Latent-structure Analysis*. Abt, Cambridge, Massachusetts.

Green, B. F. and Tukey, J. W. (1960). Complex analysis of variance: general problems, *Psychometrika*, **25**, 127–152.

Grizzle, J. E., Starner, C. F. and Koch, G. G. (1969). Analysis of categorical data by linear models, *Biometrics*, **25**, 489–504.

Gross, A. M. (1973). A robust confidence interval for symmetric, long-tailed distributions, *Proceedings of the National Academy of Sciences*, **70**, 1995–1997.

Gross, A. M. (1976). Confidence interval robustness with long-tailed symmetric distribution, *Journal of the American Statistical Association*, **71**, 409–416.

Gross, A. M. (1977). Confidence intervals for bisquare regression estimates, *Journal of the American Statistical Association*, **72**, 341–354.

Guarino, R. (1981). Robust Estimation Under a Finite Number of Alternatives; Compromise Estimates of Location. Ph.D. thesis, Department of Statistics, Princeton University.

Hampel, F. R. (1973). Robust estimation: a condensed partial survey, *Zeitschrift für Wahrscheinlichkeitstheorie und verwandte Gebiete*, **27**, 87–104.

Hampel, F. R. (1978). Modern trends in the theory of robustness, *Mathematische Operationsforschung und Statistik*, Ser. Statist., **9**, 425–442.

Harman, H. H. (1976). *Modern Factor Analysis*, 3rd edn. University of Chicago Press, Chicago.

Harshman, R. A. (1970). Foundations of the PARAFAC procedure: Models and conditions for an "explanatory" multi-modal factor analysis, *UCLA Working Paper on Phonetics*, **16**, 1–86. (Reprinted by University Microfilms, Ann Arbor, Mich., Order No. 10,085.)

Harshman, R. A. (1972). Determination and proof of minimum uniqueness conditions for PARAFAC1. *UCLA Working Paper on Phonetics*, **22**, 111–117.

Harshman, R. A. and Berenboum, S. A. (1980). Basic concepts underlying the PARAFAC–CANDECOMP three-way factor analysis model and its application to longitudinal data. In *Present and Past in Middle Life* (D. H. Erhom, P. A. Mussen, J. A. Clausen, N. Hoan and M. P. Honzik, eds). Academic Press, New York.

Harshman, R. A., Laderfoged, P. and Goldstein, I. (1977). Factor analysis of tongue

shapes, *Journal of the Acoustic Society of America*, **62**, 693–707.
Hartigan, J. A. (1975). *Clustering Algorithms*. Wiley, New York.
Heise, D. R. (1975). *Causal Analysis*. Wiley, New York.
Hill, M. O. (1974). Correspondence analysis: A neglected multivariate method, *Journal of the Royal Statistical Society* C, **23**, 340–354.
Hill, R. W. (1977). Robust Regression When There Are Outliers in the Carriers. Ph.D. thesis, Harvard University.
Hoaglin, D. C. and Welsch, R. E. (1978). The hat matrix in regression and ANOVA, *American Statistician*, **32**, 17–22.
Hoerl, A. E. and Kennard, R. W. (1970). Ridge regression: biased estimation for nonorthogonal problems, *Technometrics*, **12**, 55–67.
Hogeweg, P. (1976a). Iterative character weighting in numerical taxonomy, *Computers in Biology and Medicine*, **6**, 199–211.
Hogeweg, P. (1976b). *Topics in Biological Pattern Analysis*. Proefschrift (D.Sc. thesis) Utrecht. (Incorporates, *inter alia*, Hogeweg 1976a, 1974 papers by Hogeweg and Hesper and by Koek-Norman and Hogeweg and a 1975 paper by Hogeweg and Koek-Norman.)
Huber, P. J. (1973). Robust regression: asymptotics, conjectures and Monte Carlo, *Annals of Statistics*, **1**, 799–821.
Huber, P. J. (1979). Robust smoothing. In *Robustness in Statistics* (R. L. Launer and G. N. Wilkinson, eds). Academic Press, New York.
Hubert, L. J. (1974). Some applications of graph theory to clustering, *Psychometrika*, **39**, 283–309.
Hunter, J. E. (1980). Factor analysis. In *Multivariate Techniques in Human Communications Research* (P. R. Monge and J. N. Capella, eds). Academic Press, New York.
Jaeckel, L. A. (1972). Estimating regression coefficients by minimizing the dispersion of residuals, *Annals of Mathematical Statistics*, **43**, 1449–1458.
Jöreskog, K. G. and Sörbom, D. (1979). *Advances in Factor Analysis and Structural Equation Models*. Abt, Cambridge, Massachusetts.
Kafadar, K. (1979). Robust Confidence Intervals for the One- and Two-sample Problems. Ph.D. thesis, Princeton University. (Covered in large part by Technical Reports 151, 152, 153 and 154, Series 2, Department of Statistics, Princeton University.)
Kleiner, B. and Hartigan, J. A. (1981). Representing points in many dimensions by trees and castles, *Journal of the American Statistical Association*, **76**, 260–276.
Kleiner, B., Martin, R. D. and Thomson, D. J. (1979). Robust estimation of power spectra, *Journal of the Royal Statistical Society* B, **41**, 313–338 (discussion, 338–351).
Koch, G. G., Johnson, W. D. and Tolley, H. D. (1972). A linear models approach to the analysis of survival and extent of disease in multidimensional contingency tables, *Journal of the American Statistical Association*, **67**, 783–796.
Koch, G. G., Landis, J. R., Freeman, J. L., Freeman, D. H., Jr. and Lehnen, R. G. (1977). A general methodology for the analysis of experiments with repeated measurement of categorical data, *Biometrics*, **33**, 133–158.
Koek-Norman, J., Hogeweg, P., Van Maanen, W. H. M. and Ter Welle, B. J. H. (1979). Wood anatomy of the Blakeeae (Melastomataceae), *Acta Botanica Neerlandica*, **28**, 21–43.
Koenker, R. W. (1978). Regression quantiles, *Econometrica*, **46**, 33–50.
Kruskal, J. B. (1977). The relationship between multidimensional scaling and

clustering. In *Classification and Clustering* (J. Van Ryzin, ed.). Academic Press, New York.

Kruskal, J. B. (1978). Factor analysis and principal components: bilinear methods. In *International Encyclopedia of Statistics* (W. H. Kruskal and J. M. Tanur, eds). Free Press, New York.

Kruskal, J. B. and Wish, M. (1978). *Multidimensional Scaling*. Sage, Beverly Hills.

Larsen, W. and McCleary, S. J. (1972). The use of partial residual plots in regression analysis, *Technometrics*, **14**, 781–790.

Larsen, W. A., McGill, R. and Tukey, J. W. (1978). Variations of box plots, *American Statistician*, **32**, 12–16.

Lawley, D. N. and Maxwell, A. E. (1971). *Factor Analysis as a Statistical Method*, 2nd edn. Butterworth, London.

Leighty, R. D. (ed.) (1977). Data extraction and classification from film, *Proceedings of the Society of Photo-optical Instrumentation Engineers*, **117**, 1–142.

Lenth, R. V. (1977). Robust splines, *Communications in Statistics*, **A6**, 847–854.

Lii, K. S., Rosenblatt, M. and Van Atta, C. (1976). Bispectral measurements in turbulence, *Journal of Fluid Mechanics*, **77**, 45–62.

Lininger, L., Gail, M. H., Green, S. B. and Byar, D. P. (1979). Comparison of four tests for equality of survival curves in the presence of stratification and censoring, *Biometrika*, **66**, 419–428.

McNeil, D. R. and Tukey, J. W. (1975). Higher-order diagnosis of two-way tables, illustrated on two sets of demographic empirical distributions, *Biometrics*, **31**, 487–510.

Mallows, C. L. (1973). Some comments on C_p, *Technometrics*, **15**, 661–676.

Mallows, C. L. (1980). Some theory of nonlinear smoothers, *Annals of Statistics*, **8**, 695–715.

Mantel, N. (1981). Calculation of scores for a Wilcoxon generalization applicable to data subject to arbitrary right censorship, *American Statistician*, **35**, 244–247.

Mantel, N. and Haenszel, W. (1959). Statistical aspects of the analysis of data from retrospective studies of disease, *Journal of the National Cancer Institute*, **22**, 719–748.

Martin, R. D. (1979). Robust estimation for time series. In *Robustness in Statistics* (R. L. Launer and G. N. Wilkinson, eds). Academic Press, New York.

Miller, R. G. (1974a). The jackknife—a review, *Biometrika*, **61**, 1–15.

Miller, R. G. (1974b). An unbalanced jackknife, *Annals of Statistics*, **2**, 880–891.

Monkhouse, F. J. and Wilkinson, H. R. (1963). *Maps and Diagrams*. Methuen, London; Harper and Row, New York.

Mosteller, F. and Tukey, J. W. (1977). *Data Analysis and Regression*. Addison-Wesley, Reading, Massachusetts.

Mulaik, S. A. (1972). *The Foundations of Factor Analysis*. McGraw-Hill, New York.

Nelder, J. A. and Wedderburn, R. W. M. (1972). Generalized linear models, *Journal of the Royal Statistical Society* A, **135**, 370–384.

Olmstead, P. S. and Tukey, J. W. (1947). A corner test for association, *Annals of Mathematical Statistics*, **18**, 495–513.

Pearson, E. S. (1956). Some aspects of the geometry of statistics, *Journal of the Royal Statistical Society* A, **119**, 125–146 (discussion, 146–149).

Peto, R. and Peto, J. (1972). Asymptotically efficient rank invariant test procedures, *Journal of the Royal Statistical Society* A, **135**, 185–198 (discussion, 199–206).

Pregibon, D. (1979). Data Analytic Methods for Generalized Linear Models. Ph.D. thesis, University of Toronto.

Prentice, R. L. and Marek, P. (1979). A qualitative discrepancy between censored data rank tests, *Biometrics*, **35**, 861–867.
Rosenfeld, A. (ed.) (1976). *Digital Picture Analysis*. Springer, Berlin.
Shepard, R. N. (1978). The circumplex and related topological manifolds in the study of perception. In *Theory Construction and Data Analysis in the Behavioral Sciences* (S. Shye, ed.), 29–89. Jossey-Bass, San Francisco.
Shepard, R. N. and Carroll, J. D. (1966). Parametric representation of nonlinear data structures. In *Multivariate Analysis I, Proceedings* (*Proceedings of the International Symposium 1965*) (P. R. Krishinaiah, ed.), 561–592. Academic Press, New York.
Shye, S. (ed.) (1978). *Theory Construction and Data Analysis in the Behavioral Sciences*. Jossey-Bass, San Francisco.
Smith, G. and Campbell, F. (1980). A critique of some ridge regression methods, *Journal of the American Statistical Association*, **75**, 74–81 (discussion, 81–103).
Sneath, P. H. A. (1977a). Cluster significance tests and their relation to measures of overlap. In *Premieres Journées Internationales Analyse des Données et Informatique 1977*, 15–36. IRIA, Le Chesnay.
Sneath, P. H. A. (1977b). A method for testing the distinctness of clusters: a test of the disjunction of two clusters in Euclidean space as measured by their overlap, *Mathematical Geology*, **9**, 123–143.
Sneath, P. H. A. (1979). The sampling distribution of the W statistic of disjunction for the arbitrary division of a random rectangular distribution, *Mathematical Geology*, **11**, 423–429.
Sneath, P. H. A. and Sokal, R. R. (1973). *Numerical Taxonomy*. Freeman, San Francisco.
Sonquist, J. A. and Morgan, J. N. (1964). *The Detection of Interaction Effects*. Monograph No. 35, Survey Research Center, Institute for Social Research, University of Michigan. (Sixth edn., 1970.)
Sonquist, J. A., Baker, E. L. and Morgan, J. N. (1973). *Searching for Structure*. Ann Arbor Survey Research Center, Institute for Social Research, University of Michigan.
Spjøtvoll, E. (1977). Alternatives to plotting C_p in multiple regression, *Biometrika*, **64**, 1–8.
Steiger, J. H. and Schönemann, P. H. (1978). A history of factor indeterminacy. In *Theory Construction and Data Analysis in the Behavioral Sciences* (S. Shye, ed.), 136–178. Jossey-Bass, San Francisco.
Tarone, R. E. and Ware, J. (1977). On distribution-free tests for equality of survival distributions, *Biometrika*, **64**, 156–160.
Theil, H. (1965). The analysis of disturbances in regression analysis, *Journal of the American Statistical Association*, **60**, 1067–1079.
Theil, H. (1968). A simplification of the BLUS procedure for analyzing regression disturbances, *American Statistical Association*, **63**, 242–251.
Thompson, B. J. (ed.) (1978). *International Optical Computing Conference*. IEEE, New York.
Tucker, L. R. (1964). The extension of factor analysis to three-dimensional matrices. In *Contributions to Mathematical Psychology* (N. Frederiksen and H. Galliksen, eds). Holt, Rinehart and Winston, New York.
Tucker, L. R. (1972). Relations between multidimensional scaling and three-mode factor analysis, *Psychometrika*, **28**, 333–367.
Tukey, J. W. (1949). One degree of freedom for nonadditivity, *Biometrics*, **5**, 232–242.
Tukey, J. W. (1954). Causation, regression, and path analysis. In *Statistics and*

Mathematics in Biology (O. Kempthorne and J. Lush, eds), 35–66. Iowa State College Press, Ames.

Tukey, J. W. (1962). The future of data analysis, *Annals of Mathematical Statistics*, **33**, 1–67.

Tukey, J. W. (1974a). Lecture Notes for Statistics, 411, Princeton University.

Tukey, J. W. (1974b). Named and faceless values, *Sankhya*, **36A**, 125–176.

Tukey, J. W. (1977). *Exploratory Data Analysis*. Addison-Wesley, Reading, Massachusetts.

Tukey, J. W. (1979). Introduction to the dilemmas and difficulties of regression. Unpublished.

Tukey, J. W. (1980). Can we predict where "time series" should go next. In *Directions in Time Series* (D. R. Brillinger and T. C. Tiao, eds), *Proceedings of the IMS Special Topics Meeting on Time Series Analysis*, Iowa State University, Ames, Iowa. 1–31, IMS, Hayward, California.

Tukey, P. A. and Tukey, J. W. (1981). Methods for Direct and Indirect Graphic Display for Data Sets in 3 and more Dimensions. Aide-memoires for lectures I, II, III, Sheffield, England, March 1980. In *Interpreting Multivariate Data* (V. Barnett, ed.): Chapter 10: Preparation; Prechosen Sequences of Views, 189–213; Chapter 11: Data-Driven View Selection; Agglomeration and Sharpening, 215–243; Chapter 12: Summarization; Smoothing; Supplemented Views, 245–275. Wiley, New York.

Wachter, K. W. (1978). The strong limits of random matrix spectra for simple matrics of independent elements, *Annals of Probability*, **6**, 1–18.

Wahba, G. (1975). Interpolating splines methods for density estimation. I. Equispaced knots, *Annals of Statistics*, **3**, 30–48.

Wahba, G. (1979). How to smooth curves and surfaces with splines and cross-validation. *Proceedings of the 24th Conference on Design of Experiments*. US Army Research Office, Report 79-2. Also T.R. 555, Statistics Department, University of Wisconsin-Madison.

Wainer, H. and Thissen, D. (1981). Graphical data analysis. In *Annual Review of Psychology* (M. R. Rosenzweig and L. W. Porter, eds), 191–241. Annual Reviews, Palo Alto, California.

Wold, H. (1966). Nonlinear estimation by iterative least squares procedures. In *Festschrift for Jerzy Neyman* (F. N. David, ed.), 411–444. Wiley, New York.

Wood, F. S. (1973). The use of individual effects and residuals in fitting equations to data, *Technometrics*, **15**, 677–695.

Wright, S. (1960). Path coefficients and path regressions: Alternative or Complementary concepts?, *Biometrics*, **16**, 189–202.

8
COMPUTER INTENSIVE METHODS IN STATISTICS

Bradley Efron

Abstract

Four non-parametric methods, the jackknife, the delta method, the bootstrap and the smoothed bootstrap, are compared for estimating the standard deviation of the correlation coefficient. The numerical results motivate some speculation on the future of statistical theory.

1. Introduction

The Academy has invited us to speak on the theme "Recent Advances in Statistics". The time is ripe for such a discussion. After a decade of drift, following the postwar heyday of Neyman–Wald decision theory, statistics seems to be gathering momentum for another move forward. The line of advance involves the electronic computer, which is now not only very fast, but very cheap.

Most of our commonly used statistical methods were developed during a period when computation was slow and expensive. I am thinking now not only of such obvious examples as analysis of variance and chi-square goodness of fit, but of the entire parametric theory developed in this century, running say from Fisher's original papers of the 1920s to Lehmann's (1959) influential book on hypothesis testing. The honest answer to "Why assume

Dedicated to the Academy of Sciences of Lisbon on the occasion of their bicentenary.

normality?" is most often "Because then we can compute the answer". One would expect, and hope, that the ability to compute more ambitious answers would lead to more ambitious theories.

Consider, for example, a simple analysis of variance situation. We wish to test the null hypothesis that a data vector $\mathbf{X} = (X_1, X_2, \ldots, X_n)$ has mean vector $\mathbf{\eta} = \mathbf{0}$, versus the alternative that $\mathbf{\eta}$ lies in $\mathscr{L}$, a predetermined subspace of Euclidean n-space, containing $(1, 1, \ldots, 1)$, the main diagonal. The standard approach compares the angle A between $\mathbf{X}$ and $\mathscr{L}$ with a standard number A_0, and rejects the null hypothesis if A is smaller than A_0. The crucial constant A_0 is determined by normal theory, which is the only distribution for the components X_i which allows its theoretical calculation (see Efron, 1969).

Suppose though that the scientist's preferred null hypothesis is the one-sided exponential $f(x) = e^{-(x-1)}$, $x > -1$, rather than $f(x) = \exp(-x^2/2)/\sqrt{(2\pi)}$. These days there is nothing to stop the statistician from going to the computer and calculating, by Monte Carlo methods, the constant A_0 appropriate for this situation. Essentially the statistician would be replacing a universal table of constants, that of the "F-distribution", with a special table constructed to fit the situation at hand. Most of us do not use printed tables of the exponential or logarithm any more because it is easier for our calculators to compute the function in each particular case. I am suggesting the same fate for the F-tables at a more profound level of replacement.

We can carry the analysis of variance example still further away from parametric theory. Consider the residual vector $\mathbf{R} = (R_1, R_2, \ldots, R_n)$ (which is the projection of $\mathbf{X}$ into the space orthogonal to $\mathscr{L}$) and the empirical distribution $\hat{F}$ putting mass $1/n$ at each value of R_i, $i = 1, 2, \ldots, n$. The statistician can ascertain the constant A_0 by Monte Carlo sampling from $\hat{F}$. In other words, if $\mathbf{X}^* = (X_1^*, X_2^*, \ldots, X_n^*)$ has $X_i^* \stackrel{\text{IID}}{\sim} \hat{F}$, $i = 1, \ldots, n$, A_0 is that value such that A^*, the angle between $\mathbf{X}^*$ and $\mathscr{L}$, is less than A_0 5% (or whatever significance level is desired) of the time.

If this suggestion seems bizarre, it is worth noting how close it remains to standard normal theory. Instead of taking F to be a non-parametric estimate of distribution for the components X_i, normal theory takes $\hat{F} \sim \mathscr{N}(0, \hat{\sigma}^2)$, where $\hat{\sigma}^2 = \Sigma_{i=1}^n R_i^2/n$ (or, equivalently for the problem at hand, $\hat{\sigma}^2 = \Sigma R_i^2/(n-p)$, p the dimension of $\mathscr{L}$). The method for determining A_0, from this point on, is exactly the same as that described in the previous paragraph. Of course normal theory allows A_0 to be determined mathematically rather than by Monte Carlo, which is both its virtue and limitation. Freedman (1975) uses a close cousin of the non-parametric method to successfully analyze a complicated data set.

The remarks so far seem to indicate a statistical world in which mathematical theory is replaced by brute computation. In fact, computational advances precipitate new theory. An example from the last century is matrix theory and multidimensional geometry, which led, 50 years later, to Fisher's development of the analysis of variance. Efron (1979b) discusses several current areas of theoretical development which were impractical before the modern computer era: robust regression, the EM algorithm for maximum likelihood estimation from incomplete data sets, Cox's partial likelihood for censored data, cross-validation, the jackknife and the bootstrap. Here we take up the latter two topics again. The discussion, which is limited to a very special situation, is intended to illustrate how a more computer-based theory might function in practice, and what new theoretical problems of its own it might raise.

2. A simple problem

We will consider the following very special problem. Fourteen bivariate points $x_1, x_2, x_3, \ldots, x_{14}$ independently drawn from the same bivariate distribution F are observed, and the Pearson sample correlation coefficient

$$\hat{\rho} = \frac{\sum_{i=1}^{14}(x_{i1}-\bar{x}_1)(x_{i2}-\bar{x}_2)}{\sqrt{\left[\sum_{i=1}^{14}(x_{i1}-\bar{x}_1)^2 \sum_{i=1}^{14}(x_{i2}-\bar{x}_2)^2\right]}}$$

computed, where $\bar{x}_1 = \sum_{i=1}^{14} x_{i1}/14$, $\bar{x}_2 = \sum_{i=1}^{14} x_{i2}/14$. It is desired to estimate the standard deviation $\mathrm{SD}(\hat{\rho})$. Five different standard deviation estimates will be investigated. A brief description of the five follows, listed in order of increasing computational difficulty.

(1) The normal theory estimate

$$\widehat{\mathrm{SD}}_1 = (1-\hat{\rho}^2)/\sqrt{14} \qquad (2.1)$$

(Kendall and Stuart, 1958, p. 236).

(2) Tukey's jackknife estimate. Let $\hat{\rho}_{(i)}$ be the value of $\hat{\rho}$ calculated from the 13 data points $x_1, x_2, \ldots, x_{i-1}, x_{i+1}, \ldots, x_{14}$, the full data set with x_i removed, and define $\hat{\rho}_{(\cdot)} = \sum_{i=1}^{14} \hat{\rho}_{(i)}/14$. Then

$$\widehat{\mathrm{SD}}_2 = \sqrt{\left[\frac{13}{14}\sum_{i=1}^{14}(\hat{\rho}_{(i)}-\hat{\rho}_{(\cdot)})^2\right]} \qquad (2.2)$$

(Miller, 1974).

(3) The delta method, or infinitesimal jackknife. Let

$$\mu_{gh} = E_F(X_1 - E_F X_1)^g (X_2 - E_F X_2)^h,$$

the ghth mixed central moment of the random point $X = (X_1, X_2)$ under the true distribution F. By using Taylor series expansions, $\hat{\rho}$ can be approximated by a polynomial in the sample moments of $x_1, x_2, \ldots, x_{14}$. The standard deviation of this polynomial is expressed in terms of the μ_{gh} (Cramer, 1946, p. 359), and then estimated by substitution of the sample central moments $\hat{\mu}_{gh}$ to give the standard deviation estimate

$$\hat{SD}_3 = \sqrt{\left[\frac{\hat{\rho}^2}{56}\left(\frac{\hat{\mu}_{40}}{\hat{\mu}_{20}^2} + \frac{\hat{\mu}_{04}}{\hat{\mu}_{02}^2} + \frac{2\hat{\mu}_{22}}{\hat{\mu}_{20}\hat{\mu}_{02}} + \frac{4\hat{\mu}_{22}}{\hat{\mu}_{11}^2} - \frac{4\hat{\mu}_{31}}{\hat{\mu}_{11}\hat{\mu}_{20}} - \frac{4\hat{\mu}_{13}}{\hat{\mu}_{11}\hat{\mu}_{02}}\right)\right]}. \quad (2.3)$$

It turns out that delta method standard deviation estimates can always be calculated in a simpler manner, almost identical to that of the jackknife. (A proof will appear in a forthcoming paper by the author.) If $\hat{F}$ represents the empirical distribution putting mass $1/14$ at each point $x_1, x_2, \ldots, x_{14}$, let $\hat{F}_{(i)}^\varepsilon$ be the distribution putting mass $(1-\varepsilon)/14$ at $x_1, x_2, \ldots, x_{i-1}, x_{i+1}, \ldots, x_{14}$ and mass $(1+13\varepsilon)/14$ at x_i; also let $\hat{\rho}_{(i)}^\varepsilon$ be the Pearson correlation coefficient corresponding to $\hat{F}_{(i)}^\varepsilon$. Then

$$\hat{SD}_3 = \lim_{\varepsilon \to 0} \sqrt{\left[\sum_{i=1}^{14}\left(\frac{\hat{\rho}_{(i)}^\varepsilon - \hat{\rho}}{\varepsilon}\right)^2\right]/14}. \quad (2.3a)$$

This last expression is Jaeckel's "infinitesimal jackknife" estimate of standard deviation (Miller, 1974, p. 10). Notice that taking $\varepsilon = -1/13$ gives, almost, the jackknife estimate $\hat{SD}_2$. In practice $\hat{SD}_3$ can be calculated on the computer by using some sufficiently small value of ε. (The value $\varepsilon = 0.001$ was used in the simulation study of Section 3.) The advantage of expression (2.3a) over (2.3) is that the sometimes tedious algebra of the delta method is avoided. Expression (2.3a) also appears in the literature as the root mean square of the empirical influence curve (Huber, 1977, p. 25).

(4) *The bootstrap.* The sample correlation coefficient $\hat{\rho}$ can be thought of as $\hat{\rho} = \rho(\hat{F})$, where $\rho(\cdot)$ is the functional mapping a bivariate distribution function F into its correlation coefficient ρ. Another functional of interest is $\sigma(F)$, the standard deviation of $\hat{\rho}$ obtained from data $X_1, X_2, \ldots, X_{14} \stackrel{\text{IID}}{\sim} F$. The bootstrap estimate of standard deviation, Efron (1979a), is simply

$$\hat{SD}_4 = \sigma(\hat{F}). \quad (2.4)$$

In other words $\hat{SD}_4$ is the non-parametric maximum likelihood estimate of $\sigma(F)$. It is related to $\sigma(F)$ exactly as $\hat{\rho} = \rho(\hat{F})$ is related to $\rho = \rho(F)$.

Except for unusually favorable situations, not including the present case, $\hat{SD}_4$ must be computed by a Monte Carlo algorithm:

(i) Construct the empirical distribution $\hat{F}$ putting mass $1/14$ at each point $x_1, x_2, \ldots, x_{14}$.
(ii) With $\hat{F}$ fixed, draw a random sample of size 14 from $\hat{F}$, say, $X_1^*, X_2^*, \ldots, X_n^* \stackrel{\text{IID}}{\sim} F$ (i.e. $X_1^*, \ldots, X_{14}^*$ is a simple random sample drawn with replacement from $\{x_1, x_2, \ldots, x_{14}\}$).

(iii) Let $\hat{F}^*$ be the empirical distribution of $X_1^*, X_2^*, \ldots, X_{14}^*$ and $\hat{\rho}^* = \rho(\hat{F}^*)$. Repeat step (ii) independently N times, obtaining bootstrap replications $\hat{\rho}^{*1}, \hat{\rho}^{*2}, \ldots, \hat{\rho}^{*N}$. Approximate $\hat{SD}_4$ by

$$\hat{SD}_4 = \sqrt{\left\{\sum_{j=1}^{N} [\hat{\rho}^{*j} - \hat{\rho}^{*\cdot}]^2 / (N-1)\right\}}, \tag{2.4a}$$

where $\hat{\rho}^{*\cdot} = \sum_{j=1}^{N} \hat{\rho}^{*j}/N$. The values $N = 128$ and $N = 512$ were used in the simulation study of Section 3.

(5) Smoothed bootstrap. The bootstrap algorithm defined above starts from $\hat{F}$, the non-parametric maximum likelihood estimate for the true F. The statistician may be willing to assume that F, whatever it is, is reasonably smooth. In this case the bootstrap algorithm might better be started from a smoothed version of $\hat{F}$, say the convolution of $\hat{F}$ with some smooth distribution $\hat{W}$. In the simulation study of Section 3, $\hat{W}$ was taken to be a uniform distribution over a rhombus $\mathscr{R}$, where $\mathscr{R}$ was selected so that $\hat{W}$ had covariance matrix 0.25 times $\hat{\Sigma}$, the sample covariance matrix of $x_1, x_2, \ldots, x_{14}$. The points $X_1^*, X_2^*, \ldots, X_{14}^*$ were then selected independently from $\hat{F} * \hat{W}$, and $\hat{SD}_5$ calculated as in (2.4a).

It would be more natural to take $\hat{W}$ bivariate normal. However, this would be self-serving in the context of Section 3, since the simulation study selected the original points $x_1, x_2, \ldots, x_{14}$ themselves from a bivariate normal distribution.

Notice that $\hat{F} * \hat{W}$ has the same correlation coefficient as $\hat{F}$, namely $\hat{\rho}$. If the covariance matrix of $\hat{W}$ were not taken proportional to $\hat{\Sigma}$ this would not be true, and a bias would be introduced into the smoothed bootstrap procedure.

3. A simulation study

The study involved 200 trials‡ of the following situations

$$X_1, X_2, \ldots, X_{14} \stackrel{\text{IID}}{\sim} \mathscr{N}_2\left[\begin{pmatrix} 0 \\ 0 \end{pmatrix}, \begin{pmatrix} 1 & 0.5 \\ 0.5 & 1 \end{pmatrix}\right] \tag{3.1}$$

In this case it can be shown that $\hat{\rho}$ has expected value 0.486, standard deviation 0.221,

$$\hat{\rho} \sim (0.486, 0.221^2), \tag{3.2}$$

(Johnson and Kotz, 1970, p. 225).

---

‡ In what follows, "trial" refers to a new selection of the data points $X_1 = x_1, X_2 = x_2, \ldots, X_{14} = x_{14}$; "replication" refers to the bootstrap selection of $X_1^*, X_2^*, \ldots, X_{14}^*$, with $x_1, x_2, \ldots, x_{14}$ held fixed, as described in Section 2.

Table 1. A Comparison of Different Methods for Estimating SD($\hat{\rho}$)[a]

	Summary statistics, 200 trials			
	AVE	SD	CV	$\sqrt{(MSE)}$
1. Normal parametric	0.192*	0.050	0.26	0.058
2. Jackknife	0.223	0.085	0.38	0.085
3. Delta method (infinitesimal jackknife	0.175**	0.058	0.33	0.074
4. Bootstrap				
$N = 128$	0.206	0.066	0.32	0.068
$N = 512$	0.206	0.063	0.31	0.065
5. Smoothed bootstrap (rhomboid window, $N = 128$)	0.205	0.061	0.30	0.063

*Relative bias $\geqslant 0.10$.
**Relative bias $\geqslant 0.20$.
[a] $\hat{\rho}$ is the sample correlation coefficient for $x_1, x_2, \ldots, x_{14}$ generated from (3.1). The true value is SD($\hat{\rho}$) = 0.221.

Table 1 shows summary statistics for the 200 trials, for each of the methods described in Section 2. For example, the 200 values of $\widehat{SD}_1 = (1-\hat{\rho})^2/\sqrt{14}$ averaged 0.192, with sample standard deviation 0.050. The sample coefficient of variation was $0.050/0.192 = 0.26$, and the root mean square error, from the true value 0.221, equaled $[(0.221 - 0.192)^2 + (199/200)(0.050)^2]^{1/2} = 0.058$.

One asterisk indicates a relative bias $\geqslant 0.10$, e.g.

$$\left| \frac{(0.192 - 0.221)}{0.221} \right| = 0.13;$$

two asterisks indicate relative bias $\geqslant 0.20$. This notation reflects my opinion that biases smaller than 10% are negligible in this context. In any case, it should be noticed that the bias situation would look considerably different if the summary statistics were given in terms of variances rather than standard deviations. The true variance is Var($\hat{\rho}$) = $0.221^2 = 0.0488$. The averages, over 200 trials, of the estimated variances were 0.0394 (normal parametric), 0.0569 (jackknife), 0.0340 (delta method), 0.0468 (bootstrap 128), 0.0464 (bootstrap 512) and 0.0457 (smoothed bootstrap). The reader is referred to Efron and Stein (1978) for a proof that the jackknife estimate of variance is always biased upward in a certain specific sense. Among the non-parametric estimates of standard deviation, there is noticeably improved performance as we move in the direction of increasing computation, from the top toward the bottom of the table. The smoothed bootstrap has root mean square error less than 75% of that of the jackknife. Of course the normal parametric method is better still, but that is no surprise since the simulation involved only normal distributions.

8 COMPUTER INTENSIVE METHODS IN STATISTICS

Table 2 A Comparison of Different Methods for Estimating SD$(\tanh^{-1}\hat\rho)$[a]

	Summary statistics, 200 trials			
	AVE	SD	CV	MSE
1. Normal parametric $(=1/\sqrt{11} = 0.302)$	0.302	0	0	0.003
2. Jackknife	0.314	0.090	0.29	0.090
3. Delta method (infinitesimal jackknife)	0.244*	0.052	0.21	0.076
4. Bootstrap				
$N = 128$	0.301	0.065	0.22	0.065
$N = 512$	0.301	0.062	0.21	0.062
5. Smoothed bootstrap (rhomboid window, $N=128$)	0.298	0.055	0.19	0.058

*Relative bias ≥ 0.10.
[a] $\hat\rho$ is the sample correlation coefficient for $x_1, x_2, \ldots, x_{14}$ generated from (3.1). The true value is SD$(\tanh^{-1}\hat\rho) = 0.299$.

Table 2 is the same as Table 1, except that the quantity of interest is SD$(\tanh^{-1}\hat\rho)$ rather than SD$(\hat\rho)$. Fisher's transformation

$$\tanh^{-1}\hat\rho = \tfrac{1}{2}\log(1+\hat\rho)/(1-\hat\rho)$$

is variance stabilizing, that is it gives a (nearly) constant standard deviation, $1/\sqrt{11} = 0.302$, when sampling from a bivariate normal distribution. Once again the non-parametric estimates of standard deviation improve as we go from the jackknife to the smoothed bootstrap.

4. Discussion

Tables 1 and 2 raise many theoretical questions, most of which have not yet been solved.

(1) How much computation should one invest in a given situation? For estimating either SD$(\hat\rho)$ or SD$(\tanh^{-1}\hat\rho)$, it was definitely worthwhile to go from the jackknife, which requires 14 recalculations of $\hat\rho$, to the bootstrap with $N = 128$ replications. Multiplying N by 4, to $N = 512$, gave little additional improvement. In the present case it can be shown that further increase of N would be pointless, at least for the estimation of standard deviations. ($N = \infty$ would give SD $= 0.062$ for the bootstrap in Table 1, and SD $= 0.061$ for the bootstrap in Table 2.) In a real situation, as opposed to a Monte Carlo simulation, it is more difficult to choose N, though standard calculations give a rough idea of when to stop.

(2) What is the best way to invest computational effort? In my experience

the bootstrap nearly always outperforms the jackknife. Sometimes though the improvement is slight, and not worth the additional computational expense. There are many other resampling plans which have been suggested, see Remark I of Efron (1979a). All of these are closely related to the bootstrap. (For example, the jackknife can be explained as a delta method approximation to the bootstrap, as in Section 5 of Efron (1979a).) However, they may have different, perhaps better, properties in a given situation. The bootstrap, because of its maximum likelihood derivation, enjoys some theoretical advantages, which so far have been verified in my modest comparative studies, but the issue is by no means settled.

(3) What is the relation with parametric methods? The parametric method described in Section 2 is itself a version of the bootstrap. The only conceptual difference is that at step (i) of the bootstrap algorithm, $\hat{F}$ is taken to be the parametric maximum likelihood estimation of F, i.e. the bivariate normal with first and second moments matching those of the sample, rather than the non-parametric MLE. Of course the computation of the standard deviation estimate can then be carried out theoretically, rather than by Monte Carlo.

What would we do in a practical situation, where typically we would have no strong grounds for either believing or disbelieving the bivariate normal model? One approach is to use the smoothed bootstrap, sampling from $\hat{F} * \hat{W}$ where $\hat{W} \sim \mathcal{N}_2(0, c\hat{\Sigma})$, trying different values of c. The choice $c = 0$ gives the unsmoothed bootstrap, while $c = \infty$ gives the normal parametric estimate of standard deviation. Once again, there is no theory to rely on. See Sections 3 of Efron (1979a, b).

(4) How should confidence intervals be determined? Computing a standard deviation is a rough substitute for the more ambitious problem of assigning a two-sided confidence interval. Tukey's original suggestion was to use the jackknife and the standard t-table to assign confidence intervals of the form $\hat{\rho} \pm t_\alpha \cdot \hat{SD}_2$. However, there is no theoretical support for this approach, at least not for small enough sample sizes to make the t intervals much different than the normal intervals $\hat{\rho} \pm z_\alpha \cdot \hat{SD}_2$. Efron (1979c) has suggested a method of constructing confidence intervals using the entire bootstrap distribution $\hat{\rho}^{*1}, \hat{\rho}^{*2}, \ldots, \hat{\rho}^{*N}$. This method requires N to be larger than that for estimating the standard deviation, on the order of $N = 1000$. A theoretical argument involving transformations supports this method, but also suggests other approaches.

5. Conclusions

The next 20 years should be exciting ones for statisticians. My prediction is for a partial replacement of parametric models, and the accompanying mathematical calculations we have become used to, by computer intensive methods. These methods will replace "theory from a book", typified by t-tables and F-tables, by "theory from scratch", generated anew by the computer for each new data analysis problem. Theoreticians will not be put out of work though. Statistical computer methods themselves are perfectly worthy objects of mathematical attention. Their analysis, comparison and refinement is a formidable prospect for the statistical theorists of the late twentieth century.

References

Cramer, H. (1946). *Mathematical Methods of Statistics*. Princeton University Press, New Jersey.

Efron, B. (1969). Student's t-test under symmetry conditions, *Journal of the American Statistical Association*, **64**, 1278–1302.

Efron, B. (1979a). Bootstrap methods: another look at the jackknife, *Annals of Statistics*, **7**, 1–26.

Efron, B. (1979b). Computers and the theory of statistics: thinking the unthinkable, *SIAM Review*, **21**, 460–480.

Efron, B. (1979c). Censored Data and the Bootstrap. Technical Report No. 53, Department of Statistics, Stanford University.

Efron, B. and Stein, C. (1978). Bias of the Jackknife Estimate of Variance. Technical Report No. 40, Department of Statistics, Stanford University.

Freedman, D. A. (1975). Estimating Standard Errors in a Complex Sample Survey. Technical Report, University of California, Berkeley.

Huber, P. J. (1977). *Robust Statistical Procedures*. Society for Industrial and Applied Mathematics, Philadelphia.

Johnson, N. and Kotz, S. (1970). *Continuous Univariate Distributions*, Vol. 2. Wiley, New York.

Kendall, M. G. and Stuart, A. (1958). *The Advanced Theory of Statistics*, Vol. I. Hafner, New York.

Lehmann, E. L. (1959). *Testing Statistical Hypotheses*. Wiley, New York.

Miller, R. G. (1974). The jackknife—a review, *Biometrika*, **61**, 1–15.

9
CURRENT ISSUES IN ROBUST STATISTICS

Peter J. Huber

Abstract

The published account of the "Princeton robustness year" (Andrews et al., 1972) settled, if any, the problem of robustly estimating a location parameter in samples of size 10 or larger. The unpublished seminars and informal discussions ranged much farther, and strikingly, most of the seminal ideas tossed up 10 years ago still are burning issues today.

The present paper critically reviews the main achievements in multiparameter regression estimation when the independent variables are known and error free (giving some hints about how to deal with leverage points) and touches upon multiparameter regression in problems when there are gross errors in the independent variables. In what concerns computation of robust regression estimates, two competing algorithms are available: one approach iteratively manipulates the weights, the other the residuals.

Since the Princeton robustness year, there have been advances in other areas, in particular in covariance estimation and, recently, in time series problems. On the other hand, not much has been achieved in the general one-parameter estimation problem, in analysis of variance, and Bayesian robustness has stagnated.

Adaptive estimates, which at one time seemed to be the ultimate goal of robust estimation, nowadays should better be classified under the general heading of efficient non-parametric procedures than under robustness: in robustness the emphasis rests much more on safety than on efficiency.

1. Introduction

Ten years ago, in the summer of 1970, the so-called "Princeton robustness year" began. While the published account of that cooperative endeavor

(Andrews et al., 1972) is exclusively concerned with location estimation, the unpublished seminars and informal discussions ranged much farther, and strikingly, the catalogue of burning issues does not look that much different now from what it looked in 1971. True, there have been advances, in particular with regard to regression and covariance estimation, but not much else has been definitively settled in the meantime, and few of the intriguing ideas on time series and on analysis of variance and the like, tossed up in our seminars 10 years ago (mostly by Tukey) have been followed up in a serious fashion. Thus, any discussion of current issues must refer back to the Princeton robustness year 1970/71. It seems appropriate to begin by summarizing the state of affairs in location estimation after the winding up of that study.

2. Point estimation

If the Princeton study settled one thing, it was the problem of robustly estimating a location parameter in samples of size 10 or larger. We did not come up with "the" ultimate robust estimate; instead, we realized that there was no such thing. Clearly, we were faced with vague, vector valued personal utility functions, and the choice of the estimate would always depend both on the peculiarities of the application and the idiosyncrasies of the statistician. On the other hand, we had identified (or confirmed) several families of good robust estimates, and had learned how to summarily assess their qualities and weaknesses in a simple fashion, namely with the aid of Hampel's notions of the breakdown point (the largest fraction of gross outliers the estimate can cope with) and of the influence function (the suitably standardized influence toward the value of the estimate of one additional observation in a large sample) (Hampel, 1968, 1974).

In particular, we had learned to hand-tailor M-estimates to one's particular needs. None of the other types of estimates can be fine-tuned so fully and easily. L-estimates (like the trimmed means) lost much ground, despite their simplicity, because of their poor breakdown point when the trimming proportion is low. R-estimates (like the Hodges–Lehmann estimate, the median of pairwise means) remained attractive only because they allow natural, distribution-free confidence intervals through the help of the associated rank tests.

Thus, for samples of size 10 and larger, we had essentially reached a consensus: the estimates of choice are M-estimates, with $S = \text{MAD}$ used as an auxiliary estimate of scale. Here, MAD (or median absolute deviation) denotes the median $\text{med}[|x_i - \text{med}(x_i)|]$ of the absolute deviations from the median of the sample. In other words, the location estimate T would be determined from an implicit equation

$$\sum \psi\left(\frac{x_i - T}{S}\right) = 0,$$

where ψ is either a monotone function like Huber's minimax ψ:

$$\psi(x) = \max[-c, \min(c, x)],$$

or a redescending one like Hampel's piecewise linear function

$$\begin{aligned}
\psi(x) &= x & &\text{for } 0 \leqslant x \leqslant a, \\
&= a & &\text{for } a \leqslant x \leqslant b, \\
&= a\frac{c-x}{c-b} & &\text{for } b \leqslant x \leqslant c, \\
&= 0 & &\text{for } x \geqslant c, \\
&= -\psi(-x) & &\text{for } x < 0,
\end{aligned}$$

or Andrews' sinewave, or Tukey's biweight

$$\begin{aligned}
\psi(x) &= x\left[1 - \left(\frac{x}{c}\right)^2\right]^2 & &\text{for } |x| \leqslant c, \\
&= 0 & &\text{otherwise,}
\end{aligned}$$

or Collins' (1976) hyperbolic tangent.

The purpose of these redescending ψ is to reduce the influence of very far out observations to 0, and thus to gain some efficiency (10–15%) for very-long-tailed distributions, while sacrificing only little efficiency at the normal. The price to be paid is a slight increase of the minimax risk (about 5–10%). However, redescending estimates are rather sensitive to a wrong scale, so some care is indicated, especially for small samples (that is, in terms of Hampel's estimate, one should then increase both b and c, while keeping $a/(c-b) \leqslant 0.5$).

Somewhat less than a consensus has been reached for sample sizes below 10. This author believes that the same estimates work adequately down to sample size 5, and if the scale can be borrowed from elsewhere, even all the way down to 1. Of course, for sample sizes below 10 it becomes progressively more difficult to assess the accuracy of the estimate from information contained in the sample itself.

3. Studentizing and confidence intervals

For sample sizes 10 or larger, the more customary robust M-estimates of location are very nearly normally distributed, and their asymptotic variance can be estimated either by the jackknife, or preferably, by using a sample version of the asymptotic variance formula:

$$D^2 = \frac{1}{n} \frac{[1/(n-1)]\Sigma\psi[(x_i-T)/S]^2 S^2}{\{(1/n)\Sigma\psi'[(x_i-T)/S]\}^2}.$$

This formula is essentially equivalent to the jackknife, but it has the advantage that it is able to give a danger signal: it becomes unreliable (together with the jackknife) when the denominator contains too many small or negative terms. D may be used as a surrogate for the estimated standard deviation of the location estimate T. For sample sizes above 10 or maybe 15, the normal approximation should be good enough, so that $(T-2D, T+2D)$ can be used as an approximate 95% confidence interval.

It is tempting to conjecture that for sample sizes 10 or lower, the distribution of $(T-\theta)/D$ can be approximated by a t-distribution, but with how many degrees of freedom? The literature contains several recommendations with regard to this number of degrees, but in the author's opinion all of them, including his own, are worthless. Note that the tails of the distribution of $(T-\theta)/D$ depend both on the tails of T and on the density of D near 0, and neither of them is known for small sample sizes. Both depend considerably on the tail behavior of the unknown true underlying error distribution. It is easy to conjecture that the appropriate number of degrees of freedom is somewhat lower than the classical $n-1$, but by how much, is anybody's guess.

This, and the related problem of finding approximate 95% confidence intervals, have been important open problems for some years now. In the author's opinion, both a theoretical investigation, elucidating the qualitative dependence between the error distribution and the distribution of $(T-\theta)/D$, and a simulation study, nailing down the quantitative aspects, would be needed. The recent development of small sample asymptotics (cf. Hampel, 1973; Field and Hampel, 1980) for the first time gives some hope toward a successful two-pronged attack of the just-described kind.

In passing, we note that small sample confidence intervals are difficult to find without detailed prior information about the tails of the underlying error law. For example, for sample sizes ≤ 4, the traditional 95% t-interval for the mean always contains the sample in its entirety, cf. Berengut (1981). For sample sizes 6–8, the smallest distribution-free 95% interval for the median is the range $(x_{(1)}, x_{(n)})$. In other words, small sample confidence intervals need to know more about the tails of the underlying error law than the sample can possibly provide.

4. The general one-parameter estimation problem

The asymptotic minimax approach of Huber (1964) depends in a crucial way on underlying symmetries (e.g. on translation invariance) in order to extend the parametrization throughout an entire neighborhood of the parametric model. There is an alternative approach by Hampel (1968, 1974), staying strictly at the parametric model, minimizing the asymptotic variance there, subject to Fisher consistency and subject to an upper bound on the gross error sensitivity (i.e. on the supremum of the absolute value of the influence function).

For M-estimation, Hampel's approach works well and actually yields the same type of minimax solution as Huber's in the cases accessible to the latter. Hampel's idea still needs some refinements (cf. Huber, 1981, section 11.1), but in principle it applies to any one-parameter estimation problem. Somewhat surprisingly, it appears that it has been generalized and applied to multiparameter regression problems (see below) before it was ever put to practical use in its original form!

An awkward feature of this approach is, of course, that it remains strictly at the idealized model and only allows infinitesimal perturbations. This is a serious conceptual inadequacy, which, incidentally, causes problems and may lead to nonsensical and misleading results with R- and L-estimates.

A slightly more satisfactory approach is based on tests between shrinking neighborhoods of hypotheses approaching each other at the same rate $n^{-1/2}$. It was first explored by Huber–Carol (1970), and has more recently been taken up by Rieder (1978), Bickel (1979) and others. Technically, this approach is much more complicated, and it gives more insight into what is going on, but it leads to qualitatively the same results as Hampel's, where they overlap, cf. Rieder (1978), Theorem 3.7.

5. Multiparameter regression

The past few years have seen much action in multiparameter regression problems. Most of the investigations have been devoted to achieving robustness with regard to errors in the "dependent" observations y, assuming the "independent" variables (the matrix X) to be known and error free. Very few hard results have been obtained with regard to the problem of how to deal with gross errors in the independent variables.

With regard to computation of M-estimates of regression, most of the dust seems to have settled now, and the picture has become satisfactorily clear.

5.1. Error-free independent variables

A main difficulty is caused by so-called leverage points: observations which, thanks to their position in the design space, have an overriding influence on the fit. In particular, they can be recognized by the fact that they determine their own fitted values in a very tight way.

In an ordinary least squares regression problem

$$y = X\theta + \text{errors},$$

where $\theta = (\theta_1, \ldots, \theta_p)^T$ is the parameter vector to be estimated, and $y = (y_1, \ldots, y_n)^T$ is the vector of observations, the fitted values are given by

$$\hat{y} = X(X^TX)^{-1}X^Ty.$$

The matrix $H = X(X^TX)^{-1}X^T$ is called "hat matrix" (it puts the hat on y). Being a projection matrix, it has p eigenvalues 1 and $n-p$ eigenvalues 0; its diagonal elements satisfy $0 \leq h_{ii} \leq 1$ and

$$\max h_{ii} \geq \operatorname{ave} h_{ii} = \frac{p}{n}.$$

By definition, all points with large h_{ii} are called leverage points.

As a very simple example, take regression through the origin, with abscissae given by

$$x_i = a^i, \quad i = 0, 1, 2, \ldots, n.$$

If $a \geq 3$, the regression line is for all practical purposes determined by the last pair (x_n, y_n). The hat matrix has elements

$$h_{il} = \frac{a^2 - 1}{a^{2n+2} - 1} a^{i+l},$$

so in particular

$$h_{nn} > 1 - a^{-2}.$$

If a diagonal element h_{ii} is large, say $h_{ii} \geq 0.9$, the corresponding observation y_i for all practical purposes cannot be cross-checked on the basis of the other observations.

If the maximum diagonal element $h = \max h_{ii}$ is small, it is possible to devise an asymptotic theory of robust M-estimates of regression very similar to the well-understood theory for simple location estimates, see Huber (1973), with recent improvements by Yohai and Maronna (1979).

The issue of how one should deal with moderate to high leverage points (say $h_{ii} > 0.2$) has become quite acute. Perhaps the most important progress in this area has been that one has become aware of the problem, and that one has recognized the need for conducting alternative analyses.

The simplest recommendation is to calculate the fit twice or more times, with and without the offending leverage points. However, this does not always help. In particular, deleting one leverage point may create another one, as the above example on regression through the origin shows.

Sometimes the problem is with the model, and certain parameters (or combinations of parameters) are poorly determined by the data. Then it may help to reduce the dimensionality of the parametric model.

A more systematic way to deal with leverage points consists in downweighting observations in such a way that the maximum leverage h is cut down to a tolerable level, say 0.2 or 0.1. In the case of regression through the origin, this can be achieved very easily, namely by applying any robust location estimate, e.g. the median, to the slopes y_i/x_i. This indeed safeguards against ill influences from a gross error sitting at a leverage point. On the other hand, it increases sensitivity to the model. In this example, the influence of a minor non-linearity near the origin, or of a small error in the position of the origin, may be unduly magnified.

The idea of estimating a regression line through the origin by taking the median of the slopes has recently been generalized by Siegel and Benson (1979) to the general multiparameter regression case. For an arbitrary estimable quantity α, any p-tuple of observations in general position will give a unique determination. So define a robust estimate of α by

$$\hat{\alpha} = \underset{i_1}{\text{med}} \cdots \underset{i_{p-1}}{\text{med}} \underset{i_p}{\text{med}} \, \alpha(i_1, i_2, \ldots, i_p).$$

That is, we begin by fixing the first $p-1$ indices and taking the median over the set of all values of the pth index for which α has a well-determined value. Then we take the median over the $(p-1)$th index, and so on. This procedure yields estimates having the best possible breakdown point, namely 0.5.

The most advanced attempt to bound the overall influence of individual observations is the recent paper by Krasker and Welsch (1980), which simultaneously, iteratively bounds the influence of observations both with regard to position x and with regard to gross errors in y. The underlying idea is a direct outgrowth of Hampel's minimax approach (see Section 4).

However, this approach (and its predecessors, due to Hampel and Mallows) may be overly pessimistic. This can be seen in the case of straight line regression through the origin, where the bounded influence estimate with fixed and known scale happens to be minimax in a precise finite sample sense. It is the minimax solution for the statistician in a contamination game where nature knows the design matrix. Not surprisingly, the minimax strategy for nature then will put most of the contaminating mass on the points with the highest leverage, and the effect appears to be particularly severe in the limiting case, when the contamination tends to zero. A quantitative study of the size of the effects is still outstanding.

5.2. Robustness against errors in the carrier

In principle, all proposals for achieving robustness with regard to errors in the xs start from some robust covariance concept. This is indeed natural if we think of the geometric interpretation of fitting a straight line to a set of points in the plane: depending on the error structure in the xs and ys we are minimizing the distances between the observations and the line measured in different ways. If the xs are error free, we are measuring in the direction of the y-axis; if both x and y have the same error structure, the measurement is perpendicular to the line, and the line is actually in the direction of the principal axis of the covariance matrix of x and y.

We note that least squares regression can be described by the property that the sample cross-product between the residual vector and the columns of the design matrix is driven to 0:

$$\sum_i r_i x_{ij} = 0,$$

and that robustness with regard to gross errors in the observations y is achieved by metrically Winsorizing the residuals r_i. Thus, it is tempting to try to achieve robustness with regard to the carrier by treating the x_{ij} in a similar manner.

The conceptual flaw in this approach is apparent: we do not have an underlying model to which the xs ideally should conform, as is the case with the error process in the ys. Under the circumstances, it is not at all clear against which contingencies such a procedure does safeguard, and whether it is too pessimistic or not pessimistic enough. Presumably, one should do a careful analysis of a large sample of applied regression problems from various fields, to find out more about the structures, in particular the error structures, to be expected in the carrier.

5.3. Computation of robust regression estimates

The competing algorithms for M-estimates of regression have been winnowed down to essentially two choices: one approach iteratively manipulates the weights, the other the residuals. For some versions of robust regression, e.g. for the bounded influence regression of Krasker and Welsch (1980), the determination of the weights may become quite complicated; we shall illustrate the ideas only for a simpler version. These estimates are intended for cases where the carrier is error free.

Both algorithms split the observations into fitted values plus residuals:

$$y_i = \hat{y}_i + r_i.$$

9 CURRENT ISSUES IN ROBUST STATISTICS

The first approach uses the residuals obtained in one iteration to determine new weights

$$w_i = \frac{\psi(r_i/s_i)}{r_i/s_i}$$

to be used in the next iteration. Each iteration calculates new fitted values $\hat{y}_i$ by ordinary weighted least squares.

The second approach modifies the residuals and replaces the observations y_i by pseudo-observations

$$y_i^* = \hat{y}_i + \psi\left(\frac{r_i}{s_i}\right) s_i,$$

to be used in the next iteration to find new fitted values $\hat{y}_i$. In this case, the iterations are ordinary, unweighted least squares fits.

In the above equations, s_i is an estimate of the scale of the error in the ith observation; a widespread choice is to take all of them equal, $s_i = s$, where s might be the median absolute value of the current residuals r_i. A somewhat more sophisticated choice, introducing some safeguards against moderately high leverage points, but without actually downweighting them, is

$$s_i = \sqrt{(1-h_{ii})}s.$$

Both algorithms converge to the same fixed points, and the choice between the two primarily is one of convenience. The modified weights algorithm needs slightly fewer, but slightly more expensive iterations. Sometimes it is easier to modify weights than residuals, but just as often it is the other way around (e.g. in time series analysis and in smoothing, where unequal weights would introduce needless complications).

Convergence proofs are available for certain configurations of ψ-functions and scale estimates. Unfortunately, the case where s is the median absolute residual is not yet covered by these proofs.

For this reason, my current favorite scale estimate is found by iterating

$$s_{\text{new}}^2 := \frac{1}{(n-p)\beta} \sum_i \left[s\sqrt{(1-h_{ii})} \psi\left(\frac{r_i}{s\sqrt{(1-h_{ii})}}\right) \right]^2$$

concurrently with the modified weights or residuals iteration scheme, using my minimax ψ functions (see Section 2 and below).

This leads to a demonstrably convergent determination of the minimum of the convex function

$$\sum \rho\left(\frac{y_i - \hat{y}_i}{s\sqrt{(1-h_{ii})}}\right)(1-h_{ii})s,$$

with

$$\rho(x) = \tfrac{1}{2}x^2 + \tfrac{1}{2}\beta \qquad \text{for } |x| < c,$$
$$= c|x| - \tfrac{1}{2}c^2 + \tfrac{1}{2}\beta \qquad \text{for } |x| \geqslant c,$$

where

$$\beta = E[\psi(X)^2]$$

for a standard normal argument X, with $\psi = \rho'$. For convergence proofs see Huber (1981), Chapter 7, and Byrd and Pyne (1979).

The modified weights/residuals algorithms converge only linearly, but they reach a statistically satisfactory accuracy quite fast. There exist some more sophisticated and delicate algorithms which reach the exact solution in a finite number of steps, but the additional programming effort is considerable, while the higher accuracy will be needed only very rarely.

5.4. Redescending regression estimates?

In a regression context, redescending ψ-functions should be used only with utmost care, and I would rather advise against them. The main reason is that they are overly sensitive to a wrong scale. Since in regression one borrows strength by using one overall scale s, the resulting estimation procedures may become overly sensitive to heteroscedasticity (and the usual analysis of variance problems appear to be particularly vulnerable). Moreover, there is a serious danger that one gets caught in a local minimum. The slight increase in efficiency for long tails hardly is worth these added risks.

6. Analysis of variance

The principal difficulty with analysis of variance problems is that one tends to have potentially many parameters and a relatively low number of observations per parameter. The designs tend to be pretty well balanced, so, if we call observations with $h_{ii} > 0.2$ leverage points, it may very well happen that everything is a leverage point! This is not as absurd as it might sound at first: a single gross error anywhere may exert leverage and introduce spurious interactions.

We need to robustize the standard machinery of ANOVA. How should one robustly test for the presence of interaction, so that a few outliers can neither fool you into believing that there are interactions when there are none, nor are able to cover up real interactions by raising the overall noise level in the residuals?

If the number of observations per cell is relatively large (say $\geqslant 5$), the solution is obvious and straightforward. First, we robustly estimate individual cell effects, together with an overall asymptotic variance estimate in the style of Section 3, possibly with some higher order corrections as suggested in Huber (1973) (see also Huber, 1981, section 7.10). Then we use chi-square tests to test for the presence of interactions, or for non-zero main effects, and so on.

Problems arise if the number of observations per cell is smaller. Then we cannot start by estimating individual cell effects first, since this would leave us exposed to leverage effects of gross errors.

It should be worthwhile to think very carefully about a specific example, say a two-way design with a small number of observations per cell (1–3).

Presumably, one would start by robustly estimating main effects. Tukey (1977) for this purpose has proposed his so-called median polish, which is, essentially, L_1-regression, but any robust regression estimates might be used. This should allow us to spot anomalous cells and outlying observations much better than the classical least squares fit (which tends to reduce large residuals by smearing them into other cells).

The next step, or next higher degree of sophistication, is trickier. Presumably, one would introduce one or a few additional parameters (in the style of Tukey's one degree of freedom for non-additivity), again do a robust fit, and test against the simple additive model by a chi-square test (or, perhaps more accurately, by a slightly modified F-test with a robust variance estimate D^2 in the denominator).

7. Covariance estimation

Maronna's (1976) paper was a big step forward with regard to robust covariance matrix estimation. The advantage of his M-estimate approach is that it gives affinely invariant estimates, which automatically are positive definite. On the other hand, all affinely invariant M-estimates (and probably all affinely invariant estimates whatsoever) have a disappointingly low breakdown point, which is conjectured to be $\leqslant 1/(p+1)$, and proven to be $\leqslant 1/p$, where p is the dimension of the matrix.

There is some hope that by relaxing the invariance requirement, say to orthogonal invariance by using a robust version of principal component analysis, it might be possible to raise the breakdown point to more comfortable levels, but no hard results seem to be available yet.

8. Time series problems

The recent paper by Kleiner *et al.* (1979) shows that we finally have gained a firm grip on robust time series analysis, or at least, on the treatment of a certain class of time series problems. They achieve robustness in a straightforward way by modifying the residuals in the manner of Section 5.3. The results are striking: while in the classical approach some almost invisible outliers blot out all of the weak high frequency part of the spectrum, it comes out nicely after robustification.

9. Bayesian robustness

Curiously, Bayesian robustness has stagnated. Most Bayesians still seem to confound robustness with admissible estimation in a parametric supermodel: typically, they introduce a tail-length parameter (like the number of degrees of freedom in the t-family) which is to be estimated simultaneously with location and scale.

This was a reasonable approach 10–15 years ago, when one still was groping around for concepts and had to strain one's intuition to invent hopefully robust procedures. Now, with our improved insight, the emphasis has shifted. In the Bayesian context, the focus of robustness should have become to give guidelines on how to choose the supermodel and the prior distribution, so that the resulting procedure turns out to be demonstrably robust. It is inexcusable if this point is still left to *ad hoc* intuition in 1980. The paper by Rubin (1977) is a modest beginning of a more rational approach.

10. Exploratory/confirmatory data analysis and bootstrapping

The different standards of stringency needed in the exploratory and in the confirmatory stage of data analysis seem to have caused occasional misunderstandings. So long as one stays within the exploratory stage, quite crude surrogates for significance levels suffice to sort the suspected wheat from the probable chaff. As a consequence, new and untried methods can be put to good use in the EDA context before they become suitable for CDA.

The behavior of the classical procedures is well known, well understood and well tabulated, but for just one particular, ideal underlying situation (the IID normal case). Therefore, the practical accuracy of the classical confidence statements usually is overrated. On the other hand, since robust methods are supposed to be used not only at the exact model, their performance characteristics cannot be given with precision, even though tables for their performance at the normal model presumably would remain more accurate under deviations from the ideal case than the classical ones. In any case, if one wants to keep a reasonably high degree of reliability also under deviations from the normal distribution, bootstrap-type methods would seem to be indispensable for assessing the actual performance both in EDA and CDA.

11. Adaptive estimates

At one time, it almost seemed as if the ultimate goal of robust estimation was to find estimators achieving full asymptotic efficiency everywhere, for all sufficiently regular underlying distributions. In other words, the estimator should "adapt" to the unknown underlying distribution. This involves some very interesting and challenging problems, and is a highly attractive goal, but after much soul-searching I have reached the conclusion that adaptive estimates should better be classified under the general heading of efficient non-parametric procedures, than under robustness. The crucial point is that in robustness the emphasis rests much more on safety than on efficiency. For very large samples, where at first blush adaptive estimates look particularly attractive, the statistical variability of the estimate falls below its potential bias (caused by contamination or the like), and robustness would, therefore, suggest a move toward a relatively inefficient estimate, namely the sample median, which minimizes bias.

References

Andrews, D. F., Bickel, P. J., Hampel, F. R., Huber, P. J., Rogers, W. H. and Tukey, J. W. (1972). *Robust Estimates of Location: Survey and Advances.* Princeton University Press, Princeton.

Berengut, D. (1981). Some small sample properties of the student t-confidence intervals, *American Statistician*, **35**, 144–147.

Bickel, P. J. (1979). Talk at IMS-Meeting in Boulder, Colorado, Oct. 1979.

Byrd, R. H. and Pyne, D. A. (1979). Convergence of the Iteratively Reweighted Least Squares Algorithm for Robust Regression. Technical Report No. 313, The John Hopkins University.

Collins, J. R. (1976). Robust estimation of a location parameter in the presence of asymmetry, *Annals of Statistics*, **4**, 68–85.

Field, C. A. and Hampel, F. R. (1980). Small sample asymptotic distributions of M-estimates of location, submitted to *Biometrika*.

Hampel, F. R. (1968). Contributions to the Theory of Robust Estimation. Ph.D. thesis, University of California, Berkeley.

Hampel, F. R. (1973). Some small sample asymptotics, *Proceedings of the Prague Symposium on Asymptotic Statistics, Prague 1973.*

Hampel, F. R. (1974). The influence curve and its role in robust estimation, *Journal of the American Statistical Association*, **62**, 1179–1186.

Huber, P. J. (1964). Robust estimation of a location parameter, *Annals of Mathematical Statistics*, **35**, 73–101.

Huber, P. J. (1973). Robust regression: asymptotics, conjectures and Monte Carlo, *Annals of Statistics*, **1**, 799–821.

Huber, P. J. (1981). *Robust Statistics.* Wiley, New York.

Huber-Carol, C. (1970). Etude Asymptotique de Tests Robustes. Ph.D. thesis, Eidgen. Techn. Hochschule, Zurich.

Kleiner, B., Martin, R. D. and Thomson, D. J. (1979). Robust estimation of power spectra, *Journal of the Royal Statistical Society* B, **41**, 313–351.

Krasker, W. S. and Welsch, R. E. (1980). Efficient bounded-influence regression estimation using alternative definitions of sensitivity. To appear in *Journal of the American Statistical Association*.

Maronna, R. A. (1976). Robust M-estimators of multivariate location and scatter, *Annals of Statistics*, **4**, 51–67.

Rieder, H. (1978). A robust asymptotic testing model, *Annals of Statistics*, **6**, 1080–1094.

Rubin, H. (1977). Robust Bayesian Estimation, In: *Statistical Decision Theory and Related Topics* (S. S. Gupta and D. S. Moore, eds), Vol. II. Academic Press, New York.

Siegel, A. F. and Benson, R. H. (1979). Resistant Fitting as a Basis for Estimating Allometric Change in Animal Morphology. Technical Report No. 552, University of Wisconsin, Madison.

Tukey, J. W. (1977). *Exploratory Data Analysis*. Addison-Wesley, Reading, Massachusetts.

Yohai, V. J. and Maronna, R. A. (1979). Asymptotic behavior of M-estimators for the linear model, *Annals of Statistics*, **7**, 258–268.

10

BAYESIAN EVALUATION OF LIFE TEST SAMPLING PLANS

Richard E. Barlow and Alexander S. Wu

Abstract

Life testing has been an active research area for a long time. However, most of the past literature on life testing is based on the sampling theory paradigm. Procedures based on this paradigm are usually optimal in some sense for large sample sizes. Since it is usually difficult and expensive to obtain large sample sizes in practice, these procedures are limited in their usefulness. This motivated us to study the life testing problem from a behavioristic Bayes point of view.

Within the Bayesian paradigm, we consider selected life test sampling plans with respect to life distribution model robustness. Our objective is to obtain results useful in actually selecting life test sampling plans. Suppose we are committed to using a certain life distribution model. We may speculate, prior to observing the data, about the effect of alternative life distribution models on our estimator. Since the exponential life distribution model is the most common one in life testing, we compute the exact Bayes estimator under exponentiality and the natural conjugate prior. Then we study its robustness relative to model misspecification.

The posterior mean is the Bayes estimator with respect to squared error loss. When the model is actually misspecified, the posterior mean will in general neither be Bayesian unbiased nor minimize the expected mean squared error. We evaluate the extent of these discrepancies with respect to the sampling plan chosen, as well as for alternative classes of life distributions.

This research was supported by the Office of Naval Research under Contract N00014-75-C-0781 with the University of California. Reproduction in whole or in part is permitted for any purpose of the US Government.

1. Introduction

In this paper we consider selected life test sampling plans with respect to life distribution model robustness. Our objective is to obtain results useful in selecting life test sampling plans. To this end, let $F(\cdot|\theta)$ be a life distribution with mean life $\theta \in \Theta = [0, \infty)$. Since θ is unknown to us, we model our uncertainty about θ by letting $\tilde{\theta}$ denote a random variable with density $\pi(\theta)$. We suppose π is chosen and fixed. It is important to remark that θ and π may be considered independently of our life distribution model $F(\cdot|\theta)$ subject only to measurability restrictions. In some parts of this paper we will concentrate on the prior density

$$\pi(\theta) = \frac{b^a \theta^{-(a+1)} e^{-b/\theta}}{\Gamma(a)}, \tag{1.1}$$

where $a, b > 0$. Later, we also assume $a > 2$. This is the *inverted gamma* prior which also happens to be the natural conjugate prior when

$$F(x|\theta) = 1 - e^{-x/\theta}.$$

With this setup, suppose we choose and perform a life test experiment, $\mathscr{E}$, and observe data, D. After specifying a life distribution, $F(\cdot|\theta)$, and prior π, we may compute, $\pi(\theta|D)$, the posterior density for θ given the data, D. Let expected mean square error be our criterion for merit of an estimator of θ. Then we seek $\hat{\theta}(D)$ such that

$$E_{\pi, F}\{\tilde{\theta} - \hat{\theta}(D)\}^2$$

is minimum. Expectation is with respect to the probability measure on the sample space, determined by our sampling plan, and Θ, induced by our life distribution model $F(\cdot|\theta)$ and prior density π. It is well known that

$$\hat{\theta}(D) = E(\tilde{\theta}|D) = \int_\Theta \theta \pi(\theta|D) \, d\theta, \tag{1.2}$$

the mean of the posterior density, $\pi(\theta|D)$, minimizes expected mean squared error among estimators based on the data. $E(\tilde{\theta}|D)$ is called the *Bayes estimator* of θ with respect to squared error loss. Also, it is well known that

$$\hat{\theta}(D) = E(\tilde{\theta}|D)$$

is Bayesian unbiased; i.e.

$$E_{\pi, F}\{\hat{\theta}(D)\} = E_{\pi, F} \int_\Theta \theta \pi(\theta|D) \, d\theta = \theta_0, \tag{1.3}$$

where θ_0 is the mean of the prior density, π. That is, we expect (preposterior analysis) the posterior mean to be the same as the mean of the prior. Since we have not actually observed any data, this is perfectly consistent. Hence, given $\pi, F(\cdot|\theta)$ and our experiment $\mathscr{E}$, the mean of the posterior is Bayesian unbiased and has least expected mean squared error among estimators of θ based on the data, D.

Suppose, however, that we are committed to using a certain life distribution model; namely the exponential

$$G(x|\theta) = 1 - e^{-x/\theta}.$$

This often happens in industry. A contract may specify the life distribution model and procedures to be used to demonstrate a reliability goal. If this is the case, we may *speculate*, prior to observing data, about the effect of alternative life distribution models on our estimator. For experiment $\mathscr{E}$, let

$$\hat{\theta}_G(D) = E(\tilde{\theta}|D)$$

be our Bayes estimator for θ where the posterior is computed using the exponential life distribution model, $G(\cdot|\theta)$. If G is incorrect and $F(\cdot|\theta)$ is a more accurate model, then in general,

$$E_{\pi,F}\{\hat{\theta}_G(D)\} \neq \theta_0,$$

i.e. our estimator, based on G, is not necessarily unbiased if observations come from $F(\cdot|\theta)$. Also,

$$E_{\pi,F}\{\tilde{\theta} - \hat{\theta}_F(D)\}^2 \leq E_{\pi,F}\{\tilde{\theta} - \hat{\theta}_G(D)\}^2 \tag{1.4}$$

so that neither is $\hat{\theta}_G(D)$ best with respect to expected mean squared error. Our problem is to evaluate the extent of these discrepancies with respect to the sampling plan chosen, i.e. the number tested and the stopping rule employed, as well as for alternative classes of life distributions. We call this paper a preposterior analysis because we are in effect speculating about the posterior distribution *before* any data are actually observed. We hope to be able to recommend which sampling plan to employ when the model, say the exponential, is actually misspecified.

At the data analysis stage with the data in hand, we might pursue a Bayesian robustness study relative to our life distribution model. For example, we might consider a class of life distributions $F(\cdot|\alpha, \theta)$ indexed by mean life, θ, *and* a parameter α. For $\alpha = 1$, we might have

$$F(x|\alpha = 1, \theta) = 1 - e^{-(x/\theta)}.$$

Then, for a joint prior $\pi(\alpha, \theta)$, we compute the marginal posterior density

$$\pi^*(\theta|D) = \int \pi(\alpha, \theta|D) \, d\alpha$$

and

$$\pi(\theta|D) = \pi(\theta|\alpha = 1, D).$$

If the posterior densities, or their means, are "close" we then conclude that our procedures based on $\alpha = 1$, and our data D, are robust for the family $F(\cdot|\alpha, \theta)$ (Box and Tiao, 1972, Chapter 3). We do *not* pursue this type of robustness here. In contrast, our speculations and analysis are all *in the mind*. Our specified prior, π, is our only reference with respect to knowledge about the mean life, θ.

Our main results concern Bayesian bias of $\hat{\theta}(D)$ under selected stopping rules when we conjecture that $F(\cdot|\theta)$ may actually have decreasing mean residual life conditional on θ. See Theorems 3.1 and 4.1. Our second criterion is the expected mean squared error of our estimator when $F(\cdot|\theta)$ is *not* exponential. Results for this criterion are given in Sections 2 and 3.

Sampling plan design

For most of the situations encountered in practice, the sampling plan described in Section 4 is the best choice as it gives us control over the time span of the life test. Hence we will limit discussion on sampling plan design to this particular sampling plan. What remains to be determined are *the size of the sample put on test, n*, and *the time to terminate the test, t*.

In practical applications, we are more concerned about an estimator's expected mean squared error than we are about its Bayesian bias. Therefore we will focus our attention on the expected mean squared error as the principal criterion in designing a sampling plan.

Consider the estimator $\hat{\theta}_G(t)$ discussed in Section 4,

$$\hat{\theta}_G(t) = \frac{b+T}{a+k-1}, \qquad (1.5)$$

where k is the number of observed failures in $[0, t)$ and T is the total time on test. It is possible to calculate $E_{\pi,G}[(\hat{\theta}_G(t)-\tilde{\theta})^2]$ and discover how it varies with the sample size, n, and the termination time, t:

$$E_{\pi,G}[(\hat{\theta}_G(t)-\tilde{\theta})^2] = E_\pi\{E_G[(\hat{\theta}_G(t)-\tilde{\theta})^2|\tilde{\theta}]\}. \qquad (1.6)$$

Notice that $\hat{\theta}_G(t)$ in (1.5) above is a function of the sufficient statistic (k, T), whose conditional joint density may be obtained from a result in Bartholomew (1963) as follows:

$$P(k, T|\tilde{\theta}) = \binom{n}{k}\frac{\tilde{\theta}^{-k}}{(k-1)!}e^{-T/\tilde{\theta}} \sum_{i=0}^{k}\binom{k}{i}(-1)^i\{\max[0, T-t(n-k+i)]\}^{k-1}$$

$$\text{for } k \geqslant 1, \qquad (1.7)$$

and for $k = 0$, the probability of observing no failures in $[0, t)$ is

$$\Pr(k = 0, T = nt|\tilde{\theta}) = e^{-nt/\tilde{\theta}}. \qquad (1.8)$$

10 BAYESIAN EVALUATION OF LIFE TEST SAMPLING PLANS

Note that $(n-k)t \leqslant T \leqslant nt$ for all $1 \leqslant k \leqslant n$. From (1.7) and (1.8), we obtain

$$E_G[(\hat{\theta}_G(t) - \tilde{\theta})^2 | \tilde{\theta}] = \sum_{k=1}^{n} \int_{(n-k)t}^{nt} \left(\frac{b+T}{a+k-1} - \tilde{\theta}\right)^2 P(k, T|\tilde{\theta}) \, dT$$

$$+ \left(\frac{b+nt}{a-1} - \tilde{\theta}\right)^2 e^{-nt/\tilde{\theta}}, \tag{1.9}$$

where $P(k, T|\tilde{\theta})$ is given by (1.7). Then, by (1.6), we have

$$E_{\pi,G}[(\hat{\theta}_G(t) - \tilde{\theta})^2] = \int_0^\infty E_G[(\hat{\theta}_G(t) - \theta)^2 | \theta] \pi(\theta) \, d\theta, \tag{1.10}$$

where $E_G[(\hat{\theta}_G(t) - \theta)^2 | \theta]$ is given by (1.9), and $\pi(\theta)$ is the natural conjugate prior. The actual computation of (1.10) may be done numerically for given n, t and a, b.

Based on the accumulated knowledge and past data about the mean parameter $\tilde{\theta}$, we can determine a, b. Then we can compute $E_{\pi,G}[(\hat{\theta}_G(t) - \tilde{\theta})^2]$ in (1.10) numerically for (n, t) values being considered for the design of the sampling plan. The result of this computation may be used to choose an appropriate (n, t) value for our life test sampling plan.

The expected mean squared error for $\hat{\theta}_G(t)$ when $F(\cdot|\theta)$ is a Weibull distribution with $\alpha > 1$, should be greater than the expected mean squared error when $\alpha = 1$ and $k \doteq nF(t|\theta) \ll n$; i.e. when $F(t|\theta) \ll 1$. Hence if t is

Table 1. Complete Sampling Plan

	Sample size	
	Finite n	$n \to \infty$
Bayes estimator under exponentiality and natural conjugate prior	$\hat{\theta}_G(D) = (1-w)\theta_0 + w\bar{x}$ where $\theta_0 \sim$ prior mean $\bar{x}$ = sample mean $w = n/(a+n-1)$	$\hat{\theta}_G(D) \to \bar{x}$ $\bar{x}$ = sample mean
Bayes risk w.r.t. squared error loss under exponentiality and natural conjugate prior	$E_{\pi,G}[\tilde{\theta} - \hat{\theta}_G(D)]^2$ $= (1-w)^2\gamma^2 + (w^2/n)E_\pi(\tilde{\theta}^2)$ where γ^2 = prior variance $w = n/(a+n-1)$	$E_{\pi,G}[\tilde{\theta} - \hat{\theta}_G(D)]^2 \to 0$
Expected mean squared error of $\hat{\theta}(D)$ under general $F(\cdot|\theta)$	$E_{\pi,F}[\tilde{\theta} - \hat{\theta}(D)]^2$ $= (1-w)^2\gamma^2 + (w^2/n)E_\pi[\text{Var}_F(X|\tilde{\theta})]$ where $w = \gamma^2/\{(1/n)E_\pi[\text{Var}_F(X|\tilde{\theta})] + \gamma^2\}$	$E_{\pi,F}[\tilde{\theta} - \hat{\theta}(D)]^2 \to 0$

Table 2. Incomplete Sampling Plans

	Sampling plan		
	Stop at kth failure ($k \geq 1$)	Stop at time t ($t > 0$)	Inverse binomial sampling plan
Bayes' estimator under exponentiality and natural conjugate prior	$\theta_G(k) = (1-w)\theta_0 + w\dfrac{T}{k}$ where $\theta_0 \sim$ prior mean $T \sim$ total time on test $w = \dfrac{k}{a+k-1}$	$\theta_G(t) = (1-w)\theta_0 + w\dfrac{T}{k}$ where $\theta_0 \sim$ prior mean $k \sim$ number of failures $T \sim$ total time on test $w = \dfrac{k}{a+k-1}$	$\theta_G(k,t) = (1-w)\theta_0 + w\dfrac{T}{k}$ where $\theta_0 \sim$ prior mean $T \sim$ total time on test $w = \dfrac{k}{a+k-1}$
Bayes' risk w.r.t. squared error loss under exponentiality and natural conjugate prior	$E_{\pi,G}[\bar{\theta} - \theta_G(k)]^2$ $= \left[(1-w)^2 + \dfrac{w^2}{k}\right]\gamma^2 + \dfrac{w^2}{k}\theta_0^2$ where $\theta_0 \sim$ prior mean $\gamma^2 \sim$ prior variance $w = \dfrac{k}{a+k-1}$	No closed form expression. May be computed numerically by use of the probability density in Bartholomew (1963).	$E_{\pi,G}[\bar{\theta} - \theta_G(k,t)]^2$ $= \left[(1-w)^2 + \dfrac{w^2}{k}\right]\gamma^2 + \dfrac{w^2}{k}\theta_0^2$ where $\theta_0 \sim$ prior mean $\gamma^2 \sim$ prior variance $w = \dfrac{k}{a+k-1}$
Expected mean squared error of $\theta_G(\cdot)$ under general $F(\cdot\|\theta)$	$E_{\pi,F}[\bar{\theta} - \theta_G(k)]^2$ $= w^2\left\{\mathrm{Var}_\pi\left[E_F\left(\dfrac{T}{k}\Big\|\bar{\theta}\right)\right]\right.$ $+ E_\pi\left[\mathrm{Var}_F\left(\dfrac{T}{k}\Big\|\bar{\theta}\right)\right]$ $- 2w\,\mathrm{Cov}_\pi\left[E_F\left(\dfrac{T}{k}\Big\|\bar{\theta}\right), \bar{\theta}\right] + \gamma^2$ $\left.+ w^2\left\{E_\pi\left[E_F\left(\dfrac{T}{k}\Big\|\bar{\theta}\right)\right] - \theta_0\right\}^2\right\}$	No closed form expression	$E_{\pi,F}[\bar{\theta} - \theta_G(k,t)]^2$ $= w^2\left\{\mathrm{Var}_\pi[E_F[Z_i\|\bar{\theta}]]\right.$ $+ \dfrac{1}{k}E_\pi[\mathrm{Var}_F(Z_i\|\bar{\theta})]$ $- 2w\,\mathrm{Cov}_\pi[E_F[Z_i\|\bar{\theta}], \bar{\theta}] + \gamma^2$ $\left.+ w^2\{E_\pi[E_F[Z_i\|\bar{\theta}]] - \theta_0\}^2\right\}$ where $Z_i \sim$ time between the $(i-1)$st and the ith actual failure
Expected mean squared error of $\theta_G(\cdot)$ under general $F(\cdot\|\theta)$ in asymptotic case	$E_{\pi,F}[\bar{\theta} - \theta_G(k)]^2$ $\to (\gamma^2 + \theta_0^2)\left(\dfrac{1}{c}H^{-1}\int_0^{(c)} \bar{H}(v)\,dv - 1\right)^2$ where $\dfrac{k}{n} \to c < 1$ ($n \to \infty$, $k \to \infty$) $H(y) \equiv F(\theta y\|\theta)$	$E_{\pi,F}[\bar{\theta}_G(t)]^2$ $\to \mathrm{Var}_\pi\left(\dfrac{1}{F(t\|\bar{\theta})}\int_0^t F(u\|\bar{\theta})\,du - \bar{\theta}\right)$ $+ \left[E_\pi\left(\dfrac{1}{F(t\|\bar{\theta})}\int_0^t F(u\|\bar{\theta})\,du\right) - \theta_0\right]^2$ as $n \to \infty$	$E_{\pi,F}[\bar{\theta} - \theta_G(k,t)]^2$ $\to \mathrm{Var}_\pi[E_F(Z_i\|\bar{\theta})]$ $- 2\,\mathrm{Cov}_\pi[E_F(Z_i\|\bar{\theta}), \bar{\theta}] + \gamma^2$ $+ \{E_\pi[E_F(Z_i\|\bar{\theta})] - \theta_0\}^2$ as $k \to \infty$

much less than θ_0, the expected mean squared error computed under exponentiality should unfortunately be too small in this case.

Tables 1 and 2 summarize results obtained in subsequent sections. Inverse binomial sampling and additional asymptotic results are discussed in Wu (1980).

2. Complete samples

In this section, we suppose n units are life tested until failure so that our data $D = (x_1, x_2, \ldots, x_n)$ consists of n observed lifetimes. Given $F(\cdot|\theta)$, these are assumed independent conditional on θ. If the random lifetime, X, has distribution

$$G(x|\theta) = 1 - e^{-x/\theta},$$

then n and $\bar{x} = (1/n)\sum_1^n x_i$ is sufficient for θ. Let π be the natural conjugate prior given by (1.1) with $a > 2$. Then, the Bayes estimator is

$$\hat{\theta}(D) = (1-w)\theta_0 + w\bar{x}, \tag{2.1}$$

where

$$\theta_0 = \frac{b}{a-1}, \qquad w = \gamma^2 \bigg/ \left\{\frac{1}{n} E \operatorname{Var}(X|\tilde{\theta}) + \gamma^2\right\}$$

and

$$\gamma^2 = \frac{b^2}{(a-1)^2(a-2)}.$$

We assume henceforth, that $a > 2$ so that $\gamma^2 < \infty$. For the prior given by (1.1), $w = n/(a+n-1)$.

It can be shown (Bühlmann, 1970) that (2.1) is actually the least squares linear, in observations, approximation to the mean of the posterior density for *any* $F(\cdot|\theta)$ and *any* prior π with mean θ_0 and variance γ^2. For the exponential distribution $G(\cdot|\theta)$, (2.1) is the mean of the posterior *only* if π is the natural conjugate prior (Diaconis and Ylvisaker, 1979).

Clearly, for *any* $F(\cdot|\theta)$ and π,

$$E_{\pi,F}\{\hat{\theta}(D)\} = (1-w)\theta_0 + w\theta_0 = \theta_0 \tag{2.2}$$

so that (2.1) is always Bayes unbiased. Also, for any $F(\cdot|\theta)$ and prior π,

$$E_{\pi,F}\{\tilde{\theta} - \hat{\theta}(D)\}^2 = (1-w)^2\gamma^2 + \frac{w^2}{n} E \operatorname{Var}(X|\tilde{\theta}), \tag{2.3}$$

where X has distribution $F(\cdot|\theta)$. Since

$$\lim_{n\to\infty} w = \lim_{n\to\infty} \frac{\gamma^2}{(1/n)E\,\mathrm{Var}(X|\tilde{\theta})+\gamma^2} = 1,$$

it follows that the limiting expected mean squared error is zero for all $F(\cdot|\theta)$ and π. Hence, sample size n can be chosen to achieve any specified expected mean squared error given $F(\cdot|\theta)$ and π.

Let $\mathrm{CV}(X|\theta) = [\sqrt{\mathrm{Var}(X|\theta)}]/\theta$, the *coefficient of variation* of X. Let Y have distribution $G(x|\theta) = 1 - e^{-x/\theta}$.

Theorem 2.1. If $\mathrm{CV}(X|\theta) \leq (\geq) 1$ for all $\theta \in \Theta$, then

$$E_{\pi,F}\{\tilde{\theta}-\hat{\theta}(D)\}^2 \leq (\geq) E_{\pi,G}\{\tilde{\theta}-\hat{\theta}(D)\}^2, \tag{2.4}$$

where $\hat{\theta}(D)$ is given by (2.1).

PROOF. For the exponential distribution $G(\cdot|\theta)$, $\mathrm{CV}(Y|\theta) = 1$ for all θ. Hence, if $\mathrm{CV}(X|\theta) \leq (\geq) 1$, then

$$\mathrm{Var}(X|\tilde{\theta}) \leq (\geq) \tilde{\theta}^2$$

implies

$$E\,\mathrm{Var}(X|\tilde{\theta}) \leq (\geq) E\tilde{\theta}^2 = E\,\mathrm{Var}(Y|\tilde{\theta}). \tag{2.5}$$

(2.4) now follows from (2.5) and (2.3). □

From Theorem 2.1, it is clear that if $F(\cdot|\theta)$ is a member of any of the classes IFR, IFRA, NBU or NBUE (see Barlow and Proschan, 1975), then the expected mean squared error for $\hat{\theta}(D)$ is less than that for the exponential model. Hence, the sample size n can be determined assuming model exponentiality to meet an expected mean squared error criterion and it follows that the expected mean squared error will be even less if $F(\cdot|\theta)$ is in any of the previous classes. Of course, if F is DFR, DFRA, NWU or NWUE, the expected mean squared error will be greater than in the exponential case.

The *Bayes risk* is the expected mean squared error with respect to $F(\cdot|\theta)$ and π when $\hat{\theta}(D)$ is the posterior mean based on using $G(\cdot|\theta)$. From (2.4), we see easily that $\mathrm{CV}(X|\theta) \leq 1$ implies the Bayes risk under $F(\cdot|\theta)$ and π is less than the Bayes risk under G since

$$\min_a E_{\pi,F}(\tilde{\theta}-a)^2 = E_{\pi,F}\{\tilde{\theta}-E(\tilde{\theta}|D)\}^2 \leq E_{\pi,G}\{\tilde{\theta}-\hat{\theta}(D)\}^2$$

where $E(\tilde{\theta}|D)$ is computed with respect to the posterior distribution based on $F(\cdot|\theta)$.

3. Observation until the kth failure

In this section, we suppose that n units are life tested until the kth observed failure occurs, k is fixed in advance. We call this sampling plan (a). At this point testing stops. Although this sampling plan is perhaps not as useful as the plan which stops at a preassigned time t, it is easier to study. Also, results for this stopping rule suggest similar results for the case of time truncated sampling, sampling plan (b).

Our data consists of the k ordered lifetimes $D = \{x_{(1)} \leq x_{(2)} \leq \cdots \leq x_{(k)}\}$. If the random lifetime X has distribution

$$G(x|\theta) = 1 - e^{-x/\theta},$$

then the total time on test, $T = \sum_{i=1}^{k} x_{(i)} + (n-k)x_{(k)}$, is sufficient for θ. Let π be the natural conjugate (1.1). Then, the Bayes estimator is

$$\hat{\theta}(k) = (1-w)\theta_0 + w\frac{T}{k}, \tag{3.1}$$

where $w = \gamma^2 / \{(1/k) E \operatorname{Var}(X|\tilde{\theta}) + \gamma^2\}$. Since, under exponentiality and conditional on θ, T is the sum of k independent exponentials, (3.1) is actually the least squares linear, in T, approximation to the mean of the posterior density for any prior π with mean θ_0 and variance γ^2 as in Section 2.

Although (3.1) is Bayes unbiased for the exponential model and *any* prior, it is *not* necessarily unbiased for arbitrary $F(\cdot|\theta)$.

Theorem 3.1. *If $F(\cdot|\theta)$ has decreasing mean residual life conditional on θ, then*

$$E_{\pi,F}\{\hat{\theta}(k)\} \geq \theta_0. \tag{3.2}$$

The prior is the natural conjugate (1.1). Compare this with Theorem 4.1 in the next section. The inequality is reversed if $F(\cdot|\theta)$ has increasing mean residual life.

PROOF. Assume $F(\cdot|\theta)$ has decreasing mean residual life. For $k = n$,

$$\hat{\theta}(n) = \frac{a-1}{a+n-1}\theta_0 + \frac{n}{a+n-1}\bar{x}$$

and $E_{\pi,F}\hat{\theta}(n) = \theta_0$. Applying Bayes' Theorem and properties of the natural conjugate, we have for $k < n$,

$$\hat{\theta}(n) = (1-\hat{w})\hat{\theta}(k) + \hat{w}\frac{T\{x_{(k)}, \infty\}}{n-k}, \tag{3.3}$$

where $\hat{w} = (n-k)/(a+n-1)$ and $T\{x_{(k)}, \infty\}$ is the total time on test from $x_{(k)}$

to the last ordered failure time. Taking expectations, we have

$$\theta_0 = \left(\frac{a+k-1}{a+n-1}\right)E\hat{\theta}(k) + \left(\frac{n-k}{a+n-1}\right)E\theta^*\{x_{(k)}\}, \qquad (3.4)$$

where $\theta^*\{x_{(k)}\}$ is the mean residual life conditional on θ and given survival to $x_{(k)}$. By assumption, $\theta^*\{x_{(k)}\} \leq \theta$ for all $x_{(k)}$ and θ so that

$$E_{\pi,F}\theta^*\{x_{(k)}\} \leq \theta_0.$$

Hence, from (3.4), we must have

$$E_{\pi,F}\hat{\theta}(k) \geq \theta_0.$$

The proof when $F(\cdot|\theta)$ has decreasing mean residual life is similar. □

We call the class of failure distributions with decreasing mean residual life DMRL. If $F(\cdot|\theta)$ is IFR, then $F(\cdot|\theta)$ is DMRL. This is *not* true in general if IFR is replaced by IFRA. However, a useful comparison is available under star ordering ($\underset{*}{\leq}$) of distribution functions (Barlow and Proschan, 1975, Chapter 4).

REMARK. If $F \underset{*}{\leq} G$ and G is exponential, then F is IFRA.

Theorem 3.2. If $F_1(\cdot|\theta) \underset{*}{\leq} F_2(\cdot|\theta), \forall \theta \in \Theta$, then

$$E_{\pi,F_1}\{\hat{\theta}(k)\} \geq E_{\pi,F_2}\{\hat{\theta}(k)\}. \qquad (3.5)$$

$\hat{\theta}(k)$ is given by (3.1) and is calculated under G. This strengthens (3.2) relative to star ordering.

PROOF. From (3.1), $\hat{\theta}(k) = (1-w)\theta_0 + wT/k$ and

$$w = \frac{\gamma^2}{(1/n)E\,\text{Var}(x|\tilde{\theta}) + \gamma^2}$$

is fixed and

$$T = \sum_{i=1}^{k} x_{(i)} + (n-k)x_{(k)}.$$

It follows from Barlow and Proschan (1966) that

$$E_{F_1}(T|\theta) \geq E_{F_2}(T|\theta).$$

Hence, $E_{\pi,F_1}(T) \geq E_{\pi,F_2}(T)$. Applying this result to (3.1), the theorem follows. □

Under IFR assumptions, Bayesian bias actually increases with n for fixed k. We prove this and more in

Theorem 3.3. If $F(\cdot|\theta)$ is IFR, then

$$E_{\pi,F}\{\hat{\theta}(k)|n\}$$

increases in n for fixed k and *arbitrary* prior. $\hat{\theta}(k) = E(\tilde{\theta}|k, T)$ and the prior is arbitrary with mean θ_0 and variance γ^2. Note that $\hat{\theta}(k)$ is the posterior mean under G and arbitrary prior. It is *not* necessarily linear in T.

PROOF. Assume F is IFR. It can be shown that when π is arbitrary, the posterior mean under G

$$E(\tilde{\theta}|k, T)$$

is increasing in T for fixed k. In Barlow and Proschan (1966), it is shown that when $F(\cdot|\theta)$ is IFR,

$$P(T > t|k, n, \theta)$$

is increasing in n for fixed k and conditional on θ. It follows that

$$E_{\pi,F}\{\hat{\theta}(k)|n\} = E_{\pi,F}\{E(\tilde{\theta}|k, T, n)\}$$

is increasing in n. $\square$

If $F(\cdot|\theta)$ is DFR, then $E_{\pi,F}\{\hat{\theta}(k)|n\}$ decreases in n for fixed k. For the natural conjugate prior and $\hat{\theta}(k)$ given by (3.1), $E_{\pi,F}\{\hat{\theta}(n)\} = \theta_0$. It follows that if $n_1 < n_2$ and $F(\cdot|\theta)$ is IFR then

$$E_{\pi,F}\{\hat{\theta}(k)|n_2\} \geq E_{\pi,F}\{\hat{\theta}(k)|n_1\} \geq E_{\pi,F}\{\hat{\theta}(k)|k = n\} = \theta_0.$$

All this is true for fixed k and the natural conjugate prior (1.1). If $F(\cdot|\theta)$ is DFR, all inequalities are reversed.

The expected mean squared error criterion

For sampling plan (a), not only the Bayesian bias but also the expected mean squared error can be unfavorable for $\hat{\theta}(k)$, given by (3.1). In particular, if $F(x|\theta)$ has order of contact at the origin, with respect to x, greater than that of an exponential distribution $G(x|\theta)$, then

$$\lim_{n \to \infty} E_{\pi,F}\{\tilde{\theta} - \hat{\theta}(k)\}^2 = \infty$$

for fixed k. For example, this is true if

$$F(x|\theta) = \begin{cases} 0, & x < x_0, \\ 1 - \exp\left\{\dfrac{-(x-x_0)}{\theta - x_0}\right\}, & x \geq x_0, \end{cases} \quad (3.6)$$

where $\theta \geq x_0 > 0$.

Theorem 3.4. Assume either $F(x_0|\theta) = 0$ for $x_0 > 0$ or

$$\lim_{x \downarrow 0} \frac{F(x|\theta)}{x^\alpha} = c > 0, \qquad (3.7)$$

where $\alpha > 1$. That is, $F(x|\theta)$ is $0(x^\alpha)$ as $x \downarrow 0$. Then, for $\hat{\theta}(k)$, given by (3.1) and fixed k,

$$\lim_{n \to \infty} E_{\pi,F}\{\tilde{\theta} - \hat{\theta}(k)\}^2 = \infty. \qquad (3.8)$$

PROOF. It is easy to verify that

$$E_{\pi,F}\{\tilde{\theta} - \hat{\theta}(k)\}^2 = \text{Var}_{\pi,F}\{\tilde{\theta} - \hat{\theta}(k)\} + E^2_{\pi,F}\{\tilde{\theta} - \hat{\theta}(k)\}. \qquad (3.9)$$

Hence, it is sufficient to show for fixed k that

$$\lim_{n \to \infty} E^2_{\pi,F}\{\tilde{\theta} - \hat{\theta}(k)\} = \infty.$$

Since $T = nx_{(1)} + (n-1)(x_{(2)} - x_{(1)}) + \cdots + (n-k+1)(x_{(k)} - x_{(k-1)})$, it is enough to show

$$\lim_{n \to \infty} nE(X_{1n}|\theta) = \infty$$

for all θ.

Suppose $F(x_0|\theta) = 0$ and $x_0 > 0$. Then, obviously

$$\lim_{n \to \infty} nE(X_{1n}|\theta) \geq \lim_{n \to \infty} nx_0 = \infty.$$

From condition (3.7), it follows that $F(\cdot|\theta)$ is in the domain of attraction of a Weibull distribution with shape parameter $\alpha > 1$ (Barlow and Proschan, 1975, p. 241). It follows that

$$\lim_{n \to \infty} P\{(cn)^{1/\alpha} X_{1n} > x|\theta\} = \exp\{-(\lambda x)^\alpha\} \qquad \text{for all } x \geq 0,$$

where $\theta = \Gamma(1 + 1/\alpha)/\lambda$. Hence,

$$\lim_{n \to \infty} E\{(cn)^{1/\alpha} X_{1n}|\theta\} = \lim_{n \to \infty} \int_0^\infty P\{(cn)^{1/\alpha} X_{1n} > x|\theta\} dx$$

$$= \int_0^\infty \exp[-(\lambda x)^\alpha] dx = \Gamma(1 + 1/\alpha)/\lambda = \theta$$

since the convergence is uniform in x. Therefore,

$$\lim_{n \to \infty} nE(X_{1n}|\theta) = \lim_{n \to \infty} n^{1-1/\alpha} \theta/c^{1/\alpha} = \infty$$

since $\alpha > 1$ by assumption. $\square$

For general F, we have

$$E_{\pi,F}\{\tilde{\theta}-\hat{\theta}(k)\}^2 = w^2\left[\text{Var}_\pi\left\{E\left(\frac{T}{k}\Big|\tilde{\theta}\right)\right\} + E_\pi\left\{\text{Var}\left(\frac{T}{k}\Big|\tilde{\theta}\right)\right\}\right]$$
$$- 2w\,\text{Cov}_\pi\left\{E\left(\frac{T}{k}\Big|\tilde{\theta}\right),\tilde{\theta}\right\} + \text{Var}_\pi(\tilde{\theta})$$
$$+ w^2\left[E_\pi E\left(\frac{T}{k}\Big|\tilde{\theta}\right) - \theta_0\right]^2 \qquad (3.10)$$

Now let $F(x|\theta) = 1 - \exp\{-(\lambda x)^\alpha\}$ for $\alpha > 0$ and $\theta = [\Gamma(1+1/\alpha)]/\lambda$. From (3.10) and for $k = 1$, we have when $0 < \alpha < 1$,

$$\lim_{n\to\infty} E_{\pi,F}\{\tilde{\theta}-\hat{\theta}(1)\}^2 = \text{Var}_\pi(\tilde{\theta}) + w^2\theta_0^2.$$

It follows that for $k = 1$ and the Weibull distribution with $\alpha > 0$,

$$\lim_{n\to\infty} E_{\pi,F}\{\tilde{\theta}-\hat{\theta}(1)\}^2 \geq \lim_{n\to\infty} E_{\pi,G}\{\tilde{\theta}-\hat{\theta}(1)\}^2.$$

Hence, asymptotically, for $k = 1$, the expected mean squared error for the Weibull distribution is always greater than or equal, asymptotically, to the expected mean squared error in the exponential case.

Suppose $k = n = 1$, then for $\alpha > 1$,

$$E_{\pi,F}\{\tilde{\theta}-\hat{\theta}(1)\}^2 \leq E_{\pi,G}\{\tilde{\theta}-\hat{\theta}(1)\}^2.$$

Hence, for $\alpha > 1$ and $k = 1$, the expected mean squared error as a function of n is initially less than the corresponding expected mean squared error for the exponential case. It can be shown that eventually for some n it is greater and remains greater thereafter. In the limit, of course, it is infinite, Theorem 3.4.

4. Observation until time t

In this section, we suppose n units are life tested until time t. This is a common and important life test sampling plan, called sampling plan (b). If the random lifetime X has distribution

$$G(x|\theta) = 1 - e^{-x/\theta}$$

then k, the observed number of failures and $T = \sum_{i=1}^{k} x_{(i)} + (n-k)t$, the total time on test are together sufficient statistics for θ. For the natural conjugate prior (1.1), with $a > 1$, the mean of the posterior is now

$$\hat{\theta}(t) = \begin{cases} (1-w)\theta_0 + w\dfrac{T}{k}, & \text{for } k \geq 1 \\ \theta_0 + \dfrac{T}{a-1}, & \text{for } k = 0, \end{cases} \qquad (4.1)$$

where $w = k/a(a+k-1)$. Under exponentiality and the natural conjugate prior,

$$E_{\pi,G}\{\hat{\theta}(t)\} = \theta_0 \qquad (4.2)$$

by (1.3) since the posterior mean is always Bayesian unbiased when expectation is computed with respect to the model. However, in general, $E_{\pi,F}\{\hat{\theta}(t)\} \neq \theta_0$.

Theorem 4.1. If $F(\cdot|\theta)$ has decreasing mean residual life, i.e. is DMRL, then

$$E_{\pi,F}\{\hat{\theta}(t)\} \geq \theta_0, \qquad \forall t > 0. \qquad (4.3)$$

π is the natural conjugate (1.1) for the exponential. The inequality is reversed if $F(\cdot|\theta)$ has increasing mean residual life.

PROOF. Assume $F(\cdot|\theta)$ has decreasing mean residual life. If $t = \infty$, then we have complete observations and our estimator becomes

$$\hat{\theta} = (1-w)\theta_0 + w\bar{x}$$

and $E_{\pi,F}\hat{\theta} = \theta_0$ for all $F(\cdot|\theta)$ in this case, see Section 2. Now suppose k failures occur in $[0, t)$. By Bayes' Theorem and properties of the natural conjugate prior,

$$\hat{\theta} = \begin{cases} (1-\hat{w})\hat{\theta}(t) + \dfrac{\hat{w}T(t,\infty)}{n-k}, & k < n, \\ \hat{\theta}(t), & k = n, \end{cases} \qquad (4.4)$$

where $\hat{w} = (n-k)/(a+n-1)$ and $T(t,\infty)$ is the total residual life of the remaining $n-k$ units at time t.

Let $D_1 = \{k, T(t)\}$ be the observed data for the interval $[0, t)$; i.e. the number of failures, k, and the total time on test $T(t)$ in $[0, t)$. Since $F(\cdot|\theta)$ has decreasing mean residual life conditional on θ, we have

$$E_{\pi,F}\left\{\hat{w}\dfrac{T(t,\infty)}{n-k}\bigg|D_1\right\} = E_{\pi,F}\left\{\dfrac{T(t,\infty)}{a+n-1}\bigg|D_1\right\} \leq \dfrac{(n-k)}{(a+n-1)}E(\theta|D_1). \qquad (4.5)$$

Hence,

$$E_{\pi,F}\left\{\hat{w}\dfrac{T(t,\infty)}{n-k}\bigg|D_1\right\} \leq \hat{w}E(\tilde{\theta}|D_1) = \hat{w}\hat{\theta}(t).$$

Note that, given D_1, the distribution of θ is *always* updated using G. Hence, $E(\tilde{\theta}|D_1) = \hat{\theta}(t)$. Also, $\hat{w} = 0$ if $k = n$. It follows that

$$\theta_0 = E_{\pi,F}\hat{\theta} \leq E_{\pi,F}\{(1-\hat{w})\hat{\theta}(t)\} + E_{\pi,F}\{\hat{w}\hat{\theta}(t)\}$$

and

$$\theta_0 \leqslant E_{\pi, F} \hat{\theta}(t)$$

as was to be shown.

All inequalities are reversed when $F(\cdot|\theta)$ has increasing mean residual life. □

REMARKS. If $F(\cdot|\theta)$ is IFR, then it is also DMRL so that (4.3) holds in this case. Since $F(\cdot|\theta)$ IFRA does *not* in general imply $F(\cdot|\theta)$ is DMRL, we do not know at this time whether or not $F(\cdot|\theta)$ IFRA implies (4.3).

The classical sample theory approach to the exponential model and sampling plan (b) is very unsatisfactory (Bartholomew, 1957, 1963; Barlow and Proschan, 1967). The principle difficulty is that if no failures occur in $[0, t)$, the MLE is infinite. The sample distribution properties of the MLE are also highly unsatisfactory. In the Bayesian approach, the Bayes estimator (4.1) is Bayesian unbiased under exponentiality. There is no sample theory analogue to Theorem 4.1.

5. Concluding remarks

For complete sampling, there is no Bayesian bias using (2.1). Also, the expected mean squared error is finite. For distributions in the classes IFRA and/or NBUE, DFRA and/or NWUE, the preposterior expected mean squared error is less than under exponentiality. Hence, for complete sampling and $F(\cdot|\theta)$ IFRA and/or NBUE, we may proceed under the exponential model assumption. We can, by appropriate choice of sample size, in this case control the preposterior expected mean squared error.

However, for the sampling plan which stops at the kth observation, or time t, we have positive Bayesian bias if $F(\cdot|\theta)$ is DMRL. Hence, if it is necessary to use stopping rule (a) or (b), and we have strong reason to believe that $F(\cdot|\theta)$ is *not* exponential, then we should consider a Bayesian robustness study at the data analysis stage, see Section 1. The more IFRA or DFRA in the sense of star ordering (Barlow and Proschan, 1975, Chapter 4), the more compelling the reason for a robustness study at the data analysis stage if we are committed to using an exponential model based estimator.

References

Barlow, R. E. and Proschan, F. (1966). Inequalities for linear combinations of order statistics from restricted families, *Annals of Mathematical Statistics*, **37**, 1574–92.

Barlow, R. E. and Proschan, F. (1967). Exponential life test procedures when the distribution has monotone failure rate, *Journal of the American Statistical Association*, **62**, 548–560.

Barlow, R. E. and Proschan, F. (1975). *Statistical Theory of Reliability and Life Testing: Probability Models*. Holt, Rinehart and Winston, New York.

Bartholomew, D. J. (1957). A problem in life testing, *Journal of the American Statistical Association*, **52**, 350–355.

Bartholomew, D. J. (1963). The sampling distribution of an estimate arising in life testing, *Technometrics*, **5**, 361–374.

Basu, D. (1975). Statistical information and likelihood, *Sankhyā*, **37**, 1–71.

Box, G. E. P. and Tiao, G. C. (1973). *Bayesian Inference in Statistical Analysis*. Addison-Wesley, Reading, Massachusetts.

Buhlmann, H. (1970). *Mathematical Methods in Risk Theory*. Springer-Verlag, New York.

Diaconis, R. and Ylvisaker, D. (1979). Conjugate priors for exponential families, *Annals of Statistics*, **7**, 269–281.

Wu, A. S. (1980). Bayesian Evaluation of Life Test Sampling Plans. Ph.D. thesis. Operations Research Center, University of California, Berkeley.

11

THE STATISTICAL ANALYSIS OF INCOMPLETE LIFE LENGTH DATA

Benjamin Epstein

Abstract

In the statistical literature the methodology discussed in this paper is usually classified under the headings "life testing and (or) statistical reliability" and "survival analysis". The first term is associated with engineering applications, particularly in connection with the assessment of the length of life and reliability performance of components from life test and field data. The second term is associated with biomedical applications, particularly in connection with the estimation of survival probability as a function of time following the initiation of treatment, from data obtained in medical follow-up studies.

A characteristic feature of observations from life tests and medical follow-up studies is their incompleteness. The reason for this is quite simple. Life lengths and survival times cannot be observed instantaneously and it may be too costly or take too long to wait until all of the items in a sample of items on life test have failed or until all of the individuals in a follow-up study have died.

Much has been written on various aspects of the statistical analysis of incomplete life length data. In this paper we discuss some of the main developments, placing particular emphasis on the concepts underlying the statistical methodology.

Some key words and phrases: life testing, follow-up study, censored survival data, incomplete life length data, hazard rate, failure rate, force of mortality, integrated hazard function, Poisson process, exponential distribution, order statistics, total time on test statistic, Type I and Type II censoring, Weibull distribution, hazard plots, Kaplan–Meier estimate, competing risks, proportional hazards, two sample life tests, concomitant information, Cox proportional hazard regression model.

1. Introduction

In this paper we describe and discuss a wide class of statistical problems which arise in a natural way in a variety of contexts. In the literature such problems are often considered under the headings "life testing" or "analysis of survival data". The first term is commonly used in engineering applications, the second term in biomedical applications.

The feature common to all of the problems under discussion is that, for one reason or another, some (possibly all) of the observations on items or individuals in the sample under study are incomplete. That is to say we may have a sample of size n, but only on a subset of the sample (possibly empty) do we have actual life lengths or survival times. Information about the rest of the sample is incomplete (e.g. item has not yet failed, individual is still alive etc.).

There is a vast literature devoted to various aspects of the statistical analysis of incomplete life length data. It is our purpose to discuss some of the main developments and to place them in perspective.

2. How do incomplete life length data arise?

The occurrence of incomplete observations in life test and survival studies is an intrinsic and not a contrived feature of the data. Life lengths and survival times cannot, by definition, be observed instantaneously and it may well be too expensive and (or) take too long a time to wait until all of the items in our sample have failed or until all of the individuals in a follow-up study have died. We are forced by practical considerations to terminate our life test or follow-up early, either in terms of number of items failed or time spent under observation or both.

The simplest way for incomplete data to arise is to put n items on life test at the same time. The data then become available in such a way that the smallest lifetime occurs first, the second smallest second,..., and finally the largest lifetime last. In mathematical terms the data become available as the order statistics $t_{(1)} \leq t_{(2)} \leq \cdots \leq t_{(n)}$. If the life test is discontinued as soon as we have observed the first r failures, we will have the *complete* observations $t_{(1)}, t_{(2)}, \ldots, t_{(r)}$ and $(n-r)$ *incomplete* observations. What we know about the $(n-r)$ unfailed items is that they have all survived for at least a length of time $t_{(r)}$. Of course if $r = n$, there are no incomplete observations.

This is our first example of incomplete data. In the literature this is referred to as Type II censoring. We could also have decided to stop testing at some preassigned time t^*. In this case we will observe some random number r (possibly zero) of items which fail prior to t^*, i.e. we will observe r completed

11 THE STATISTICAL ANALYSIS OF INCOMPLETE LIFE LENGTH DATA

failure times $t_{(1)} \leq t_{(2)} \leq \cdots \leq t_{(r)}$ and $(n-r)$ incomplete observations. What we know about the $(n-r)$ unfailed items is that they have all survived at least a length of time t^*. In case $r = 0$, we have only incomplete observations. Stopping the life test at time t^* is referred to in the literature as Type I censoring. We could, of course, use other stopping rules such as $\min[t_{(r)}; t^*]$ for preassigned r and t^* or various sequential rules. In all cases we obtain incomplete data.

Up to this point it was tacitly assumed that all items are placed on test simultaneously and that items that fail are not replaced. Neither of these assumptions need be satisfied. We could start with n bulbs which we place in n sockets and as soon as a bulb in a given socket burns out replace it by another new bulb. Or the starting time of each item tested could be different. For instance a doctor is interested in following up patients suffering from a certain type of cancer, who come to him during some 5-year period. The patients will arrive at different times during the 5 years and the amount of time under observation will be different for each patient. Those arriving at the beginning of the 5-year period would have a much longer potential follow-up time than those arriving for treatment later. In this case if we let $x_i =$ survival time of patient number i and $t_i =$ follow-up time, then what we actually observe for each patient is $y_i = \min(x_i, t_i)$, where $y_i = x_i$ if $x_i \leq t_i$ and $y_i = t_i$ if $x_i > t_i$.

Sometimes incomplete data occur for other reasons. For instance, in a testing laboratory there may be a certain number of testing machines and we may, therefore, have to remove items on test quite arbitrarily in order to make the test equipment available for testing something else. Or patients may drop out of a treatment (or be dropped) and be lost to further follow-up. This type of withdrawal raises basic methodological questions since the reason for dropping out or being dropped cannot be ignored in analysing the data.

Another way for incomplete data to arise is in the context of competing risks. Suppose, for the moment, that we have n individuals who are subjected simultaneously to k competing risks for failure or death. Suppose that failure or death of any individual is associated with or due to one and only one of the causes. Clearly failure or death due to one of the causes preempts failure or death due to any of the other causes. In other words failure or death due to cause A_j acts as a censor for failure or death due to other causes $A_i, i \neq j$. An interesting question is: Suppose that the random variable $X_i, i = 1, 2, \ldots, k$ is the lifetime of an individual under the assumption that only risk A_i is operative. Assume further that X_i is distributed with CDF $F_i(x)$ and that we observe for n items or individuals $y_j = \min(x_{1j}, x_{2j}, \ldots, x_{kj})$, $j = 1, 2, \ldots, n$. How can we estimate the F_is from the y_js?

We could give many more examples of how incomplete life length data can occur, but for brevity we stop at this point.

3. The analysis of incomplete life length data—exponential distribution

It is useful and instructive to begin our discussion of techniques for analysing incomplete life length data by first considering the case when lifetimes are exponentially distributed. As pointed out in Epstein (1958) the exponential distribution plays a key role in life testing and reliability. It has the virtue of simplicity and relevance and serves as a benchmark and comparison distribution for situations where the assumption of exponentiality is not satisfied. A series of papers beginning with Epstein and Sobel (1953), summarized partially in Epstein (1960b, c) and in more detail in Epstein (1960d), dealt with various facets of statistical inference from life test data, when life lengths are assumed to be exponentially distributed. In what follows we give some of the main results for this distribution and indicate their wider applicability.

Formally we assume that the random variable of interest is the lifetime T, which is exponentially distributed with mean θ, i.e. has PDF

$$f(t;\theta) = \frac{e^{-t/\theta}}{\theta}, \quad t \geq 0. \tag{3.1}$$

Suppose that n items are drawn at random from the PDF (3.1), that all items are placed on test at time $t = 0$, that observations (item failure times) become available in order of increasing size, and that testing is discontinued after the first r failures have been observed (Type II censoring). Looked at statistically the life test yields the first r order statistics $t_{(1)} \leq t_{(2)} \leq \cdots \leq t_{(r)}$ and we may, for example, wish to estimate the mean life θ, given these observations.

One way of proceeding is to find the likelihood function $L(\theta; t_{(1)}, t_{(2)}, \ldots, t_{(r)})$, which we write, for brevity, as $L(\theta)$. It is easily verified that

$$L(\theta) \propto \frac{\exp(-T_r/\theta)}{\theta^r}, \tag{3.2}$$

where T_r, the "total accumulated life time of items on test until time $t_{(r)}$", is given by

$$T_r = \sum_{i=1}^{r} t_{(i)} + (n-r)t_{(r)} = \sum_{i=1}^{r-1} t_{(i)} + (n-r+1)t_{(r)}. \tag{3.3}$$

$\sum_{i=1}^{r} t_{(i)}$ is the total lifetime on test of the r items which failed (complete observations) and $(n-r)t_{(r)}$ is the total lifetime on test of the $(n-r)$ unfailed items (incomplete observations). The maximum likelihood estimate (MLE) of θ, which we denote by $\hat{\theta}$ (the value of θ maximizing $L(\theta)$), is

$$\hat{\theta} = \frac{T_r}{r}. \tag{3.4}$$

11 THE STATISTICAL ANALYSIS OF INCOMPLETE LIFE LENGTH DATA

T_r is a sufficient statistic and $\hat{\theta}$ is the minimum variance unbiased estimate (MVUE) of θ.

It is instructive to look at the structure of T_r. To do so we find it convenient to work with the r "spacings" between item failures, which we define as

$$y_1 = t_{(1)} \quad \text{and} \quad y_i = t_{(i)} - t_{(i-1)}, \quad 2 \leq i \leq r. \tag{3.5}$$

It can be verified readily that whereas $t_{(1)}, t_{(2)}, \ldots, t_{(r)}$ are dependent random variables, the y_is are independent random variables exponentially distributed with means $\theta/n-i+1$, $i = 1, 2, \ldots, r$, respectively. To normalize the spacings, so that they are not only independent but also identically distributed with common exponential distribution with mean θ, we define "normalized spacings" z_i, $i = 1, 2, \ldots, r$, where

$$z_1 = ny_1 \quad \text{and} \quad z_i = (n-i+1)y_i, \quad 2 \leq i \leq r. \tag{3.6}$$

Physically $z_1, z_2, \ldots, z_r$ are the "total lifetimes on test" in the time intervals $(0, t_{(1)}), (t_{(1)}, t_{(2)}), \ldots, (t_{(r-1)}, t_{(r)})$, respectively, and hence T_r can be written as

$$T_r = z_1 + z_2 + \cdots + z_r = T_1 + (T_2 - T_1) + \cdots + (T_r - T_{r-1}), \tag{3.7}$$

where

$$T_j = z_1 + z_2 + \cdots + z_j, \quad j = 1, 2, \ldots, r \tag{3.8}$$

is the total accumulated lifetime on test until time $t_{(j)}$.

There is a useful stochastic interpretation of the transformation from the first r observed lifetimes $t_{(1)}, t_{(2)}, \ldots, t_{(r)}$ to the first r accumulated total lifetimes on test $T_1, T_2, \ldots, T_r$. Suppose that we associate with each item on test a homogeneous Poisson process with rate $\lambda = 1/\theta$. Placing n items on test simultaneously is equivalent to superimposing n of these Poisson processes. The superposition is a Poisson process with rate $n\lambda = n/\theta$. We start observing this process at time $t = 0$. When the first item failure occurs at time $t_{(1)}$, one of the component Poisson processes is "extinguished" and the interfailure time $t_{(2)} - t_{(1)}$, the time between the first and second item failures, is generated by a Poisson process with rate $(n-1)\lambda = (n-1)/\theta$. More generally, the interfailure time, $t_{(i)} - t_{(i-1)}$, is generated by a Poisson process with occurrence rate $(n-i+1)\lambda = (n-i+1)/\theta$, corresponding to the superposition of the $(n-i+1)$ Poisson processes associated with the unfailed items. The $t_{(i)}$s can thus be thought of as being generated by a pure death process with death rate n/θ in the time interval $0 \leq t < t_{(1)}$, with rate $(n-1)/\theta$ in the time interval $t_{(1)} \leq t < t_{(2)}, \ldots$, and with rate $(n-r+1)/\theta$ in the time interval $t_{(r-1)} \leq t < t_{(r)}$. The T_is can be thought of as generated by a homogeneous Poisson process with rate $\lambda = 1/\theta$. From this fact we conclude at once that $f_{T_r}(x)$, the PDF of T_r, is r

stage Erlang and given by

$$f_{T_r}(x) = \frac{x^{r-1} e^{-x/\theta}}{\theta^r (r-1)!}, \qquad x \geq 0. \tag{3.9}$$

This, in turn, by virtue of the well-known connection between Erlang and chi square distributions, means that $2\lambda T_r = 2T_r/\theta$ is chi square distributed with $2r$ degrees of freedom (written as $\chi^2(2r)$). This basic result is used when testing hypotheses regarding θ, computing OC curves for life test plans, and computing confidence intervals for the mean life θ (or equivalently for the associated occurrence or failure rate λ). For details see references mentioned in the first paragraph of this section of the paper.

The T_is also play a role in testing for exponentiality. Under this assumption we have seen that the T_is can be looked at as successive occurrences in a homogeneous Poisson process. Therefore the random variables

$$u_i = \frac{T_i}{T_r}, \qquad i = 1, 2, \ldots, r-1, \tag{3.10}$$

are distributed as order statistics in a sample of $r-1$ observations taken independently and at random on a random variable uniformly distributed over the unit interval. Several tests for exponentiality utilizing this fact are described in Epstein (1960a). Graphs of the u_is, $i = 1, 2, \ldots, n-1$ for general life distributions appear under the name (scaled) total time on test (TTT) plots in the book by Barlow et al. (1972) and papers by Barlow and Campo (1975), Barlow and Davis (1977), Bergman (1977a, b, 1979) and Barlow (1979).

The ordered failure times are of interest in their own right. For instance, $t_{(r)}$ is the duration of a Type II censored life test. The distribution of $t_{(r)}$ is that of the rth smallest order statistic in a sample of size n from an exponential distribution. $E(t_{(r)})$ and $\text{Var}(t_{(r)})$, the expectation and variance of $t_{(r)}$, are found most readily by noting that

$$t_{(r)} = t_{(1)} + (t_{(2)} - t_{(1)}) + \cdots + (t_{(r)} - t_{(r-1)})$$
$$= \frac{z_1}{n} + \frac{z_2}{n-1} + \cdots + \frac{z_r}{n-r+1}. \tag{3.11}$$

From the fact that the z_is are IID exponentially distributed with mean θ, we obtain

$$E(t_{(r)}) = \theta \sum_{i=1}^{r} \frac{1}{n-i+1} \tag{3.12}$$

and

$$\text{Var}(t_{(r)}) = \theta^2 \sum_{i=1}^{r} \frac{1}{(n-i+1)^2}. \tag{3.13}$$

There are many uses for (3.12). One of these is to provide an alternative estimator for θ, $\tilde{\theta} = \beta_{r,n} t_{(r)}$, where $\beta_{r,n} = 1/\sum_{i=1}^{r} 1/(n-i+1)$. $\tilde{\theta}$ is based only on the rth order statistic and is highly efficient when compared with $\hat{\theta}$, which does require knowing all of the first r failure times $t_{(1)} \leqslant t_{(2)} \leqslant \cdots \leqslant t_{(r)}$. For details see Epstein (1952, 1956, 1960d). (3.12) also provides a basis for objectively designing life tests which strike a balance between the amount of time spent testing and the cost of items placed on test. Still another consequence of (3.12) is that plotting $\sum_{i=1}^{j} 1/n-i+1$ against the ordered lifetimes $t_{(j)}$ should yield points falling approximately along a straight line, if the underlying life distribution is exponential. Such plots are called cumulative hazard plots and are quite useful for visualizing important features of the life test data. Wayne Nelson of the General Electric Research and Development Laboratories has worked extensively on such plots and some of this is described in Nelson (1972). We shall have more to say about these plots when we discuss nonparametric estimation of life distributions.

Up to now we assumed Type II censoring. Suppose instead that censoring was of Type I, i.e. testing is stopped at some fixed time t^*. In this case it is easily shown that similar to (3.2) the likelihood function is of the form

$$L(\theta) \propto \frac{\exp[-T(t^*)/\theta]}{\theta^{N(t^*)}}, \qquad (3.14)$$

where $N(t^*)$ = number of items failed before time t^* and $T(t^*)$, the total accumulated lifetime on test up to time t^*, is given by

$$T(t^*) = \sum_{i=1}^{N(t^*)} t_{(i)} + [n - N(t^*)]t^*, \qquad 1 \leqslant N(t^*) \leqslant n,$$
$$= nt^*, \qquad \text{if } N(t^*) = 0, \qquad (3.15)$$

and where $t_{(1)} \leqslant t_{(2)} \leqslant \cdots \leqslant t_{(N(t^*))} < t^*$ are the failure times of the $N(t^*)$ items failing prior to t^*.

Unlike the Type II censoring situation where $2T_r/\theta \sim \chi^2(2r)$, the distribution of the sufficient statistic $T(t^*)$ for the Type I censoring case cannot be written down in closed form. This problem was investigated in some detail by Bartholomew (1963). Fortunately it turns out, in many practical situations, that knowing $N(t^*)$, the number of items failing in $(0, t^*)$, is almost as informative as knowing the actual failure times $t_{(1)} \leqslant t_{(2)} \leqslant \cdots \leqslant t_{(N(t^*))} < t^*$. We encountered this phenomenon earlier for the Type II censoring case where the estimate $\tilde{\theta}$ based only on $t_{(r)}$ is almost as good as $\hat{\theta} = T_r/r$. Similar results hold for truncated life tests of the form stop testing at $\min[t_{(r)}; t^*]$, where both r and t^* are preassigned. Details can be found in Epstein (1954a).

The key point we want to make is that whether life length data come to us from Type I or Type II censored plans, or from truncated, sequential, truncated sequential plans, the likelihood function for the failure data obtained up to time t is of the form

$$L(\theta) \propto \frac{e^{-T(t)/\theta}}{\theta^{N(t)}} \quad \text{or} \quad L(\lambda) \propto \lambda^{N(t)} e^{-\lambda T(t)}, \tag{3.16}$$

where $N(t)$ is the number of failures observed up to time t and $T(t)$ is the total accumulated lifetime on test up to time t. If life test data are available from more than one life test then the likelihood function for all of the tests combined is obtained by multiplying the individual likelihood functions. Equivalently the number of observed failures and total accumulated lifetimes on test for all of the tests combined are, respectively, the sum of the observed failures and total accumulated lifetimes for each test.

With respect to statistical analysis, any of the techniques: frequentist (often called classical or Neyman–Pearson), Bayesian, likelihood, structural, fiducial, will make use of (3.16). An equivalent way of saying this is that the total life statistic plays a fundamental role no matter how one analyses the data.

In the classical, structural or fiducial approaches one may encounter some technical difficulties, since the exact distribution of T depends on the truncation rule. However, in many cases where the exact distribution cannot be found, it is possible to provide quite satisfactory approximations. Likelihood and Bayesian methods are intrinsically free of dependence on stop rules, but this virtue is offset by other problems. For example, intervals obtained using relative likelihood procedures are subject to various interpretations. As for Bayesian procedures, (3.16) can be used directly to update the assumed prior for θ or λ to obtain a posterior. This is formally very convenient, particularly if the prior is chosen from a "nice" conjugate family. However, there still remains the nagging question of the basis for choosing a particular prior since this will greatly influence the posterior and all conclusions drawn from it, particularly if the amount of data available for (3.16) is scanty.

In the Type I censoring case it was assumed that all incomplete lifetimes are equal to the truncation time t^*. A useful and important generalization is when truncation times may vary from item to item under observation. For instance, in a medical context, the follow-up time for each patient will generally be different depending upon when observation or treatment began. Thus for n patients with potential survival times $t_i, i = 1, 2, \ldots, n$ and follow-up times $t_i^*, i = 1, 2, \ldots, n$, what we actually observe are the n random variables $t_i' = \min(t_i, t_i^*), i = 1, 2 \ldots, n$ where $t_i' = t_i$, if $t_i \leq t_i^*$ and $t_i' = t_i^*$, if $t_i > t^*$.

Under the assumption that survival times are exponentially distributed with mean θ (or equivalently can be thought of as being generated by a

Poisson process with rate $\lambda = 1/\theta$) it follows easily that the likelihood function has the form

$$L(\theta) \propto \frac{e^{-T/\theta}}{\theta^k} \quad \text{or equivalently} \quad L(\lambda) \propto \lambda^k e^{-\lambda T}, \tag{3.17}$$

where k is the number of t_is $\leqslant$ the corresponding t_i^*s and where the total time on test T is given by

$$T = \sum_{\{i:t_i \leqslant t_i^*\}} t_i + \sum_{\{i:t_i > t_i^*\}} t_i^*. \tag{3.18}$$

Again the total life statistic plays a fundamental role and whatever comments we made about the importance of (3.16) are also relevant with respect to (3.17). As indicated earlier in the Type I censoring case all of the relevant statistical information is contained in the likelihood function (3.17). Bartholomew (1963) has investigated some of the distribution properties of the total life statistic T.

4. The hazard rate and force of mortality function

In the previous section we found it convenient, when items on life test are exponentially distributed with mean θ, to associate with each item on test a homogeneous Poisson process with rate $\lambda = 1/\theta$. For the general case where lifetimes of items on test are IID random variables with CDF $F(t)$ and PDF $f(t)$, we associate with each item on test a non-homogeneous Poisson process with occurrence rate

$$\begin{aligned} h(t) &= \lim_{\Delta \to 0} \frac{P(t < T < t + \Delta \mid T > t)}{\Delta} \\ &= \lim_{\Delta \to 0} \frac{F(t+\Delta) - F(t)}{\Delta \bar{F}(t)} \\ &= \frac{f(t)}{\bar{F}(t)}. \end{aligned} \tag{4.1}$$

In engineering applications, $\bar{F}(t) = 1 - F(t)$ is known as the reliability function (in words, the probability that lifetime T exceeds t) and h(t) is known as the (conditional) hazard or failure rate. In biomedical, demographic and actuarial applications $\bar{F}(t)$ is known as the survival function and $h(t)$ is known by such names as force of mortality and (age conditional) death rate. From (4.1) $h(t)\Delta$ is approximately the probability that an item (individual) will fail (die) in the interval $(t, t+\Delta)$ given survival up to time t.

Just as $h(t)$ is computable given $f(t)$ and $F(t)$ it is also easy to find $f(t)$ and

$F(t)$ given $h(t)$. In this case (4.1) can be expressed by the differential equation,

$$\frac{d}{dt}\left\{-\ln \bar{F}(t)\right\} = h(t). \tag{4.2}$$

Under the assumption that $\bar{F}(0) = 1$, this implies that

$$\bar{F}(t) = e^{-H(t)}, \quad t \geq 0. \tag{4.3}$$

Hence

$$F(t) = 1 - e^{-H(t)}, \quad t \geq 0, \quad \text{and} \quad f(t) = h(t)e^{-H(t)}, \quad t \geq 0. \tag{4.4}$$

In (4.3) and (4.4), $H(t) = \int_0^t h(x)\,dx$, the integrated hazard function up to time t. Another way of expressing (4.3) and (4.4) is to say that if lifetime T is distributed with CDF $F(t)$, then the random variable $H(T)$ is exponentially distributed with mean one.

From a stochastic point of view a complete observation on a random variable T with lifetime CDF $F(t)$ and PDF $f(t)$ is equivalent to observing a non-homogeneous Poisson process with occurrence (hazard, failure, mortality) rate $h(t)$ until there is an occurrence. If, during the time that we observe the stochastic process, there is no occurrence we have an incomplete observation.

In the special case where lifetimes T are exponentially distributed with PDF (3.1), $h(t) = \lambda = 1/\theta \; \forall t$. As previously noted in Section 3, lifetimes which are exponentially distributed with mean life θ can be viewed stochastically as being generated by a homogeneous Poisson process with failure rate $\lambda = 1/\theta$.

If lifetimes T are Weibull distributed with scale parameter θ and shape parameter β, i.e. with CDF.

$$F(t) = 1 - \exp\left[-\left(\frac{t}{\theta}\right)^\beta\right], \quad t \geq 0, \theta > 0, \beta > 0, \tag{4.5}$$

then $h(t) = \beta t^{\beta-1}/\theta^\beta, t \geq 0$. For the case $\beta > 1 (0 < \beta < 1)$, $h(t)$ is strictly increasing (decreasing). When $\beta = 1$, the Weibull distribution reduces to the exponential and $h(t)$ is constant.

The exponential and Weibull distributions are undoubtedly the two most important parametric probability laws for modelling life length and survival data. A more general class of life length and time to failure distributions is one for which $h(t)$ is non-decreasing (non-increasing). Such distributions are called IHR(DHR), increasing (decreasing) hazard rate or IFR(DFR) increasing (decreasing) failure rate. A larger class of life length distributions which includes the IHR(DHR) class as a subset is one for which $H(t)/t$ is non-decreasing (non-increasing). Such distributions are called IHRA(DHRA), increasing (decreasing) hazard rate average, or IFRA(DFRA), increasing (decreasing) failure rate average. The Weibull distribution for $\beta > 1$ $(0 < \beta < 1)$ is IHR(DHR). The exponential distribution is a boundary distribution and is both IHR and DHR. For details about consequences that can be obtained from the assumption that $h(t)$ or $H(t)/t$ is increasing (decreasing) see Barlow and Proschan (1965, 1975).

5. The analysis of incomplete life length data—Weibull and other distributions

In Section 3 we considered some procedures for analysing incomplete life length data, when lifetimes are exponentially distributed. This case was treated in some detail because of its importance and relevance to more general situations.

After the exponential distribution it is generally agreed that the next most important life length distribution is the Weibull, given by (4.5). The first thing we would like to point out is that, if the shape parameter β is known and only the scale parameter θ is unknown, then all of the exponential procedures can be converted easily into procedures for the Weibull distribution. For instance, in the Type II censoring case, where testing is stopped after obtaining the first r ordered failure times $t_{(1)} \leqslant t_{(2)} \leqslant \cdots \leqslant t_{(r)}$ from a sample of n items on life test, we would simply work with the ordered random variables $w_{(1)} \leqslant w_{(2)} \leqslant \cdots \leqslant w_{(r)}$, where $w_{(i)} = t_{(i)}^{\beta}$. Analogous to (3.4) we would have as the MLE of θ

$$\hat{\theta} = \left(\frac{W_r}{r}\right)^{1/\beta}, \tag{5.1}$$

where

$$W_r = \sum_{i=1}^{r-1} w_{(i)} + (n-r+1)w_{(r)}. \tag{5.2}$$

W_r can be thought of as the total time on test up to the rth failure, not in real time but in transformed time. Similarly confidence statements and tests of hypotheses about θ could be made on the basis of the fact that $2W_r/\theta^{\beta} \sim \chi^2(2r)$.

An estimate of θ based on using only the elapsed test time ($t_{(r)}$ in the case of Type II censoring) can be obtained by noting that analogous to (3.12),

$$E(w_{(r)}) = E\left[\left(\frac{t_{(r)}}{\theta}\right)^{\beta}\right] = \sum_{j=1}^{r} \frac{1}{n-j+1}. \tag{5.3}$$

This yields the estimate

$$\tilde{\theta} = \frac{t_{(r)}}{\left(\sum_{j=1}^{r} \frac{1}{n-j+1}\right)^{1/\beta}}. \tag{5.4}$$

$\tilde{\theta}$ is biased, but it is an easy exercise to find an appropriate multiplying factor $\gamma_{r,n,\beta}$ such that $\theta^* = \gamma_{r,n,\beta}\tilde{\theta}$ is an unbiased estimate of θ. Approximate confidence limits for θ can be computed using the fact that $2r(t_{(r)}/\theta)^\beta/(\sum_{j=1}^{r} 1/n-j+1)$ is approximately distributed as $\chi^2(2r)$ when $r/n < 0.5$. Thus for example one can assert with approximately $100(1-\alpha)\%$ confidence that $\theta > t_{(r)}[2r/\chi_\alpha^2(2r)\sum_{j=1}^{r} 1/n-j+1]^{1/\beta}$, where $\chi_\alpha^2(2r)$ is the upper $100\alpha\%$ point of $\chi^2(2r)$.

Similar results could be obtained if the survival function is of the form

$$\bar{F}(t) = \exp[-H_1(t)/\theta], \qquad t \geq 0 \tag{5.5}$$

or

$$\bar{F}(t) = \exp[-H_2(t/\theta)], \qquad t \geq 0, \tag{5.6}$$

where $H_1(t)$ and $H_2(t)$ are known functions of t.

For CDFs of the form (5.5) in the Type II censored case, the MLE of θ is given by

$$\hat{\theta} = \frac{\sum_{i=1}^{r-1} H_1(t_{(i)}) + (n-r+1)H_1(t_{(r)})}{r} \tag{5.7}$$

and confidence limits on θ can be found easily from the fact that $2r\hat{\theta}/\theta \sim \chi^2(2r)$.

An unbiased estimate of θ based only on $t_{(r)}$ is easily shown to be

$$\tilde{\theta} = \beta_{r,n} H_1(t_{(r)}), \tag{5.8}$$

where $\beta_{r,n} = 1/\sum_{j=1}^{r} 1/n-j+1$. As in the exponential case, $\tilde{\theta}$ has high efficiency as compared to $\hat{\theta}$ for $r/n < 0.5$ and $2r\tilde{\theta}/\theta$ is approximately distributed as $\chi^2(2r)$.

When $\bar{F}(t)$ has the form (5.6) with known $H_2(t)$, we are in effect saying that the scale parameter θ has the same dimension as time. This is physically very natural and reasonable, but the analogue of (5.7) may not exist unless $H_2(t/\theta)$

can be factored into a product of a function of t and a function of θ. However an analogue to (5.8) is easy to find.

Using the fact that

$$E[H_2(t_{(r)}/\theta)] = \sum_{j=1}^{r} 1/n-j+1, \qquad (5.9)$$

it follows easily that

$$\tilde{\theta} = t_{(r)}/H_2^{-1}\left(\sum_{j=1}^{r} 1/n-j+1\right), \qquad (5.10)$$

where $H_2^{-1}(\cdot)$ is the inverse function of $H_2(\cdot)$. $\tilde{\theta}$ may be biased, but for $H_2(t)$ known it is purely a technical matter to find the factor $\gamma_{r,n}$ such that $\theta^* = \gamma_{r,n}\tilde{\theta}$ will be an unbiased estimate of θ. From the fact that $2r\beta_{r,n}H_2(t_{(r)}/\theta)$ is approximately distributed as $\chi^2(2r)$ we can assert with approximately $100(1-\alpha)\%$ confidence that $\theta > t_{(r)}/H_2^{-1}[\chi_\alpha^2(2r)/2r\beta_{r,n}]$, where $\chi_\alpha^2(2r)$ is the upper $100\alpha\%$ point of $\chi^2(2r)$. For the special case that $H_2(t) = t^\beta$, this reduces to the one sided lower $100(1-\alpha)\%$ confidence interval previously obtained for the Weibull distribution with known shape parameter β.

To sum up, what we have said, in effect, is that if we know the CDF of life lengths up to a scale parameter, then statistical procedures for analysing incomplete life length data, when life lengths are exponentially distributed, can be appropriately modified in order to obtain point estimates of and confidence intervals for the scale parameter θ.

But what happens if the life length observations depend upon two or more unknown parameters and we have incomplete life length data? This problem has been investigated in considerable detail, particularly for the two parameter exponential distribution, mixtures of two or more exponential distributions, two and three parameter Weibull distributions, Gamma distributions, extreme value distributions, normal and log-normal distributions, power and Pareto distributions etc. In essence all of the statistical techniques involve writing down the joint likelihood function of all observed life lengths, complete and incomplete, and then applying maximum likelihood techniques to find the values of the unknown parameters maximizing the likelihood function (posterior likelihood if one uses Bayesian methods). Except for the two parameter exponential distribution (see, e.g. Epstein, 1960e), nice closed form solutions for the MLEs of the unknown parameters do not exist and one must resort to iterative numerical techniques. Such considerations lie outside the scope of this paper. For further details, particularly for the Weibull distribution, see Mann et al. (1974) and Bain (1978).

6. Nonparametric analysis of incomplete life length data

Up to this point we have restricted ourselves to parametric statistical techniques for analysing incomplete life length data. In this section we consider nonparametric or distribution free methods of analysis.

In parametric models the emphasis is on estimating the unknown parameters of the model and then using these parameter estimates to estimate other quantities of interest such as quantiles of the underlying life length distribution (e.g. the lower tenth percentile, median) or the probability of surviving some preassigned time t^*. Often the parameters themselves have some intrinsic physical meaning. For example, when the underlying life distribution is one parameter exponential, then θ is the mean lifetime of the items and $\lambda = 1/\theta$ is the failure rate of the associated Poisson process. For two parameter exponential lifetime CDFs of the form, $F(t, A, \theta) = 1 - e^{-(t-A)/\theta}$, $t \geqslant A \geqslant 0, \theta > 0$, the location parameter A is the minimum life or guarantee period and $A + \theta$ is the mean lifetime of items. For Weibull distributions of the form (4.5) the scale parameter θ is the $100(1-e^{-1})\%$ point for any value of the shape parameter β. The value of β determines the behaviour of the hazard function as a function of time.

If there are valid physical or empirical reasons for choosing a particular parametric model, then this should be done, since it thus becomes possible to obtain a great deal of information from life tests which are terminated early either in time and (or) number of items failed. But there are many situations in practice when one does not have a real basis for choosing a particular parametric model and this is where nonparametric life test estimation procedures must be used.

The pioneering paper on nonparametric estimation based on incomplete life length data is by Kaplan and Meier (1958). The motivation for this paper came from medical follow-up studies where the random variable of interest is X, survival time of individuals following some treatment of a disease which the individual is suffering from. In order to assess the efficacy of the treatment one would like to estimate the CDF $F(x) = \Pr(X \leqslant x)$ or survival function $\bar{F}(x) = \Pr(X > x)$ or perhaps estimate such quantiles as the lower 10% point of $F(x)$, or median life, mean life, probability of surviving 1 year, 2 years, 5 years etc. In the usual statistical setting we would observe, let us say, N individuals having survival times $x_1, x_2, \ldots, x_N$ and then estimate $F(x)$ ($\bar{F}(x)$) by the empirical CDF (survival function) $F_N(x)$ ($\bar{F}_N(x)$) and could readily provide nonparametric estimates of the other quantities mentioned in the previous sentence. The difficulty is that we may not be able to observe each patient long enough to obtain the actual survival time. Instead patient number i will be under observation (often called follow-up time) only for a length of time t_i and what we observe are not $x_1, x_2, \ldots, x_N$ but $y_1, y_2, \ldots, y_N$,

11 THE STATISTICAL ANALYSIS OF INCOMPLETE LIFE LENGTH DATA

where $y_i = \min(x_i, t_i)$, $i = 1, 2, \ldots, N$. The main problem treated in the Kaplan–Meier paper is the estimation of $\bar{F}(x)$ from the y_is. If $y_i = x_i(t_i)$ we have a (an) complete (incomplete) observation on the ith patient. In the Kaplan–Meier paper complete (incomplete) observations are called deaths (losses), respectively. The main problem treated in the Kaplan–Meier paper is to estimate $\bar{F}(x)$ from the y_is under the assumption that the t_is are independent of the x_is.

The Kaplan–Meier product limit (PL) estimate $\hat{P}(t)$ is determined as follows. Suppose N individuals are under observation. Let $y_i, i = 1, 2, \ldots, N$, be the observed survival time on the ith individual. Next arrange the y_is in order of increasing size, $y_{(1)} \leqslant y_{(2)} \leqslant \cdots \leqslant y_{(N)}$, then

$$\hat{P}(t) = \prod_r \frac{N-r}{N-r+1}, \tag{6.1}$$

where r runs through those positive integers for which $y_{(r)} \leqslant t$ and for which $y_{(r)}$ is a complete observation. If there are no incomplete observations then $\hat{P}(t)$ becomes $\bar{F}_N(t) = n(t)/N$ where $n(t)$ is the number of individuals surviving time t.

The PL estimate, $\hat{P}(t)$, given by (6.1) is a descending step function with downward jumps occurring only at those times when a complete observation (death) occurs. If $y_{(i)}$ is an incomplete observation (loss), $\hat{P}(y_{(i)}) = \hat{P}(y_{(i-1)})$, and the only change that takes place at $y_{(i)}$ is that the number of individuals at risk (under treatment and observation) decreases from $N-i+1$ to $N-i$. If $y_{(i)}$ is a completed survival time (death occurring at time $y_{(i)}$) then $\hat{P}(y_{(i)}) = \hat{P}(y_{(i-1)})[(N-i)/(N-i+1)]$. $\hat{P}(t) = \hat{P}(y_{(i-1)})$ $\forall t$ such that $y_{(i-1)} \leqslant t < y_{(i)}$ and where $y_{(0)} = 0$. Taking $P(0) = 1$ fully defines $\hat{P}(t)$ up to the largest observed lifetime t^*. It is tacitly assumed in the foregoing that deaths and losses occur singly, but modifications can be made easily for the case when some of the $y_{(i)}$s are tied because of the simultaneous occurrence of multiple deaths and (or) losses.

Kaplan and Meier prove that, within the class of all possible life length distributions the PL estimate (6.1) maximizes the likelihood of the observed sample. They also provide approximate formulae for $V[\hat{P}(t)]$, the variance of $\hat{P}(t)$, and for $\hat{\mu}$ and $V(\hat{\mu})$, the estimated mean lifetime and its variance and establish that $\hat{P}(t)$ is a consistent estimate of the survival function $\bar{F}(t)$. For further results on PL estimates see Efron (1967), Breslow and Crowley (1974) and Meier (1975).

In view of (4.3) which relates the survival function $\bar{F}(t)$ and the integrated hazard function $H(t)$, one could use the Kaplan–Meier estimate $\hat{P}(t)$ to obtain an estimate $\hat{H}(t)$ of $H(t)$. It is given by

$$\hat{H}(t) = -\ln \hat{P}(t). \tag{6.2}$$

Another alternative estimate is $\tilde{H}(t)$, which is defined as

$$\tilde{H}(t) = \sum_r 1/N - r + 1, \qquad (6.3)$$

where r runs through those positive integers for which $y_{(r)}$ is a complete observation. Note that $\tilde{H}(t)$ is an increasing step function with positive jumps occurring only at those times when a complete observation (death) occurs. If $y_{(i)}$ is an incomplete observation (loss) $\tilde{H}(y_{(i)}) = \tilde{H}(y_{(i-1)})$ and the only change that takes place at $y_{(i)}$ is that the number of individuals at risk decreases from $N - i + 1$ to $N - i$. If $y_{(i)}$ is a completed survival time then $\tilde{H}(y_{(i)}) = \tilde{H}(y_{(i-1)}) + 1/N - i + 1$. $\tilde{H}(t) = \tilde{H}(y_{(i-1)}) \forall t$ such that $y_{(i-1)} \leq t < y_{(i)}$ and where $y_{(0)} = 0$. Taking $\tilde{H}(0) = 0$ fully defines $\tilde{H}(t)$ up to the largest observed lifetime. If there are no incomplete observations then $\tilde{H}(y_{(r)}) = \sum_{j=1}^{r} 1/N - j + 1$, a direct consequence of the fact that if $X \sim F(x)$ then $H(X) = -\ln \bar{F}(X)$ is exponentially distributed with mean 1 and hence $E[\tilde{H}(y_{(r)})] = \sum_{j=1}^{r} 1/N - j + 1$. Thus, in effect, the $\tilde{H}$ estimate is derived using the method of moments. Just as we could estimate $H(\cdot)$ from $\hat{P}(\cdot)$ so we could estimate $\bar{F}(\cdot)$ from $\tilde{H}(\cdot)$ through the relation $\tilde{P}(t) = e^{-\tilde{H}(t)}$.

What we have just sketched is the rationale for the cumulative hazard plots of Nelson (1972). Just as $\hat{P}(t)$ provides a consistent estimate of $\bar{F}(t)$ and hence of $H(t)$, so does $\tilde{H}(t)$ provide a consistent estimate of $H(t)$ and hence of $\bar{F}(t)$. Simply stated, for very large N, it really does not make any difference whether we use the PL or cumulative hazard procedures. We leave it as a very simple exercise to verify that for large N,

$$-\ln \hat{P}(t) = \sum_r \ln \frac{N-r+1}{N-r} \sim \sum_r 1/N - r + 1 = \tilde{H}(t). \qquad (6.4)$$

It may also be desirable to smooth the step function estimates $\hat{P}(t)$ and $\tilde{H}(t)$, since it is reasonable to assume that the survival and cumulative hazard functions are smooth functions of time.

Carrying out the calculations for PL estimates or cumulative hazard plots can become quite tedious for large N. In addition it may happen that precise information is not available about the time of death or loss of each individual in the follow-up study. In this case one divides the time axis into intervals, whose number and length is appropriate to the particular set of data being analysed, and then applies actuarial and life table estimation techniques. Several references to such techniques are given in the Kaplan-Meier paper. A key idea in all of these methods is that the hazard function or force of mortality can be assumed to be approximately constant in each time interval. This very reasonable assumption reduces the estimation problem within each interval to one in which we can apply the techniques of Section 3 of this paper. Thus if $t = k\tau$, where τ is the common interval length (this is assumed for

11 THE STATISTICAL ANALYSIS OF INCOMPLETE LIFE LENGTH DATA 229

convenience), then an estimate of $\bar{F}(t) = \Pr(X > t)$ is given by $\exp[-(\Sigma_{i=1}^{k} \hat{\lambda}_i)\tau]$ where $\hat{\lambda}_i$ is the estimate of the hazard or mortality rate in the interval $((i-1)\tau, i\tau)$, $i = 1, 2, \ldots, k$.

If precise information is available about times of death and loss, then $\hat{\lambda}_i = d_i/T_i$, where d_i is the number of deaths and T_i is the total observed life in the ith interval. If, however, we know only n_{i-1}, the number of individuals under observation at the start of the ith interval ($n_0 = N$) and d_i, l_i, the number of deaths and losses, respectively, in the ith interval, then λ_i would be estimated by $\lambda_i^* = -\ln[(n_{i-1} - l_i/2 - d_i)/(n_{i-1} - l_i/2)]/\tau$. λ_i^* was obtained from the "adjusted-observed estimate" for the (conditional) probability of survival through the ith interval (Kaplan and Meier 1958, p. 472).

7. Competing risks as a source for incomplete data

In the introduction we described briefly how competing risks can give rise to incomplete data. In this section we wish to discuss some aspects of the statistical analysis of such data.

We begin by considering a situation where items can be considered as being exposed to k independent competing risks, the occurrence of any one of which can result in item failure. We assume, for simplicity, that associated with risk j is a constant occurrence (failure) rate $\lambda_j, j = 1, 2, \ldots, k$. Alternatively we can think of a system composed of k black boxes operating in series, where black box j has exponential time to failure PDF $\lambda_j \exp(-\lambda_j t)$, $t \geq 0$, $j = 1, 2, \ldots, k$ and where failure of any black box causes system failure. A failure of black box j is called a type j failure. Suppose we have time to failure data on n such systems. How do we estimate λ_j from the data? What is the estimate of π_j, the probability of a failure of type j? What is the life length distribution of those systems which have failed due to cause j? What would be the life length distribution of a system if system failures were caused only by failures of type j and all other causes did not operate? What would be the life length distribution of the system if failures of type j were completely eliminated?

To answer these questions we proceed as follows. Let $t_{ij}, j = 1, 2, \ldots, k$ be the lifetime of the jth unit in the ith system if the jth unit were allowed to operate until it actually failed. The failure time that we actually obtain for the ith system is $z_i = \min(t_{i1}, t_{i2}, \ldots, t_{ik})$, $i = 1, 2, \ldots, n$. To answer the questions asked in the previous paragraph we first note that

$$\pi_j = \int_0^\infty \lambda_j \exp(-\Lambda u)\,du = \lambda_j/\Lambda, \quad j = 1, 2, \ldots, k, \quad (7.1)$$

where $\Lambda = \Sigma_{j=1}^{k} \lambda_j$ is the *system* failure rate. Incidentally, under the exponential assumption, Θ, the system mean life, is given by

$$\Theta = \frac{1}{\sum_{j=1}^{k} 1/\theta_j}, \tag{7.2}$$

where θ_j is the mean life of the jth black box.

Suppose now that C_j denotes the event that a failure of type j occurs, or that black box j failed. In that case it is easy to show that

$$\Pr(z < Z < z+dz | C_j) = \frac{\Pr(z < Z < z+dz, C_j)}{\Pr(C_j)}$$

$$= \frac{\lambda_j e^{-\Lambda z} dz}{\lambda_j/\Lambda} = \Lambda e^{-\Lambda z} dz. \tag{7.3}$$

Put simply, the life length distribution of those systems which have failed due to cause j or equivalently the life length distribution of completed box j failure times is given by the PDF $\Lambda e^{-\Lambda t}$, $t \geq 0$, $\forall j$.

Of course, if only cause j were to operate and all other causes did not operate, the system failure time PDF would become $\lambda_j \exp(-\lambda_j t)$, $t \geq 0$. Similarly if failures of type j were completely eliminated, the system failure time PDF would become $\Lambda^{(j)} \exp(-\Lambda^{(j)} t)$, $t \geq 0$, where $\Lambda^{(j)} = \Lambda - \lambda_j$.

Given the n system failure times z_i, $i = 1, 2, \ldots, n$ and the vector $\mathbf{n} = (n_1, n_2, \ldots, n_k)$ with $n = \Sigma_{j=1}^{k} n_j$, where the jth component of $\mathbf{n}$ gives the number of system failures due to cause j (or alternatively because of a failure of black box j), how do we estimate the λ_js and π_js? The easiest way of doing this is to write down $L(\lambda_1, \lambda_2, \ldots, \lambda_k | \mathbf{n}, z_i, i = 1, 2, \ldots, n)$.

It is readily verified that

$$L(\lambda_1, \lambda_2, \ldots, \lambda_k) = \prod_{j=1}^{k} \lambda_j^{n_j} \exp(-\Lambda T_n), \tag{7.4}$$

where $T_n = \Sigma_{i=1}^{n} z_i$. From (7.4) it follows that the MLEs of λ_j are given by

$$\hat{\lambda}_j = \frac{n_j}{T_n}, \quad j = 1, 2, \ldots, k. \tag{7.5}$$

The $\hat{\lambda}_j$s are biased estimates of the respective λ_js. Using the facts that $\lambda_j = \pi_j \Lambda$, that $\hat{\Lambda} = (n-1)/T_n$ is an unbiased estimate of Λ, that $\hat{\pi}_j = n_j/n$ is an unbiased estimate of π_j and that $\hat{\pi}_j$ is independent of $\hat{\Lambda}$, then

$$\tilde{\lambda}_j = \frac{n_j(n-1)}{nT_n}, \quad j = 1, 2, \ldots, k \tag{7.6}$$

is an unbiased estimate of λ_j.

It is illuminating to find $\hat{\lambda}_j$ in a different way. Suppose we focus only on what happens to box j and we divide the n observed box j lifetimes into two sets: n_j *complete* lifetimes $z_i^{(j)}$ where $i \in C_j$ (system failure is caused by the failure of black box j) and $N - n_j$ *incomplete* lifetimes $z_i^{(l)}$ where $i \in C_l$ with $l \neq j$ (system failure is caused by the failure of some black box $l \neq j$; such failures act as censorers in so far as a failure of type C_j is concerned). Then the likelihood function for λ_j can be written as

$$L(\lambda_j) \propto \lambda_j^{n_j} \exp(-\lambda_j T_n), \tag{7.7}$$

where

$$T_n = \sum_{\{i \in C_j\}} z_i^{(j)} + \sum_{l \neq j} \sum_{\{i \in C_l\}} z_i^{(l)}. \tag{7.8}$$

Of course the MLE of λ_j obtained from (7.7) is

$$\hat{\lambda}_j = \frac{n_j}{T_n}, \tag{7.9}$$

which agrees, as it must, with (7.5).

Up to now we have assumed that associated with risk C_j (or, alternatively, system failure caused by the failure of black box j) is a constant failure rate λ_j and time to failure PDF $\lambda_j \exp(-\lambda_j t)$, $t \geq 0$, $j = 1, 2, \ldots, k$. It should be pointed out that much of what we have said can be generalized to the so-called proportional hazards case, where it is assumed that the hazard function associated with the jth risk is $h_j(t) = c_j h(t)$, $j = 1, 2, \ldots, k$. This is equivalent to assuming that the associated time to failure PDF is $c_j h(t) \exp[-c_j H(t)]$, $t \geq 0$, $j = 1, 2, \ldots, k$ (which in turn is equivalent to assuming that the survival function $\bar{F}_j(t)$ associated with the jth risk is $\bar{F}_j(t) = [\bar{F}(t)]^{c_j}$, where $\bar{F}(t) = e^{-H(t)}$). For convenience, and without loss of generality, we will assume in what follows that $\sum_{j=1}^{k} c_j = 1$. This is physically quite reasonable. If we let $h_j(t)$, $j = 1, 2, \ldots, k$ be the hazard function associated with box j and let $h_S(t)$ be the system hazard function then, for a series system with independent risks, it follows that

$$h_S(t) = \sum_{j=1}^{k} h_j(t). \tag{7.10}$$

Therefore

$$\frac{\sum_{j=1}^{k} h_j(t)}{h_S(t)} = 1. \tag{7.11}$$

Assume that $\forall t$, $h_j(t) = c_j h_S(t)$, then clearly $h(t) = h_S(t)$ and the c_js satisfy the condition $\sum_{j=1}^{k} c_j = 1$.

Corresponding to (7.1) and (7.3) are

$$\pi_j = \int_0^\infty c_j h(t) e^{-H(t)} dt = c_j, \quad j = 1, 2, \ldots, k \tag{7.12}$$

and

$$\Pr(z < Z < z+dz | C_j) = c_j h(z) e^{-H(z)} dz/c_j = h(z) e^{-H(z)} dz. \tag{7.13}$$

As in the case of exponentially distributed competing risks, estimates of $\pi_j = c_j$ are given by $\hat{\pi}_j = \hat{c}_j = n_j/n$. The only thing that remains to be done is to estimate the system survival function, $\bar{F}(t)$, or the system integrated hazard function, $H(t)$. It follows from (7.13) that all of the system failure times, independent of cause, can be used for this purpose. If all n system failure times are observed, one could estimate $\bar{F}(t)$ by the empirical survival function, $\bar{F}_n(t)$, or alternatively estimate $H(t)$ by the empirical cumulative hazard function $H_n(t)$, which is a step function with positive jump $1/n-i+1$ at the ith smallest failure time $t_{(i)}, i = 1, 2, \ldots, n$.

It has been assumed, for simplicity, that we are able to observe all of the system failure times. If some of them are not available because of censoring, it should be obvious how to modify the various estimation procedures.

It should also be pointed out that the competing risks model involving constant failure rates λ_j for the jth risk, $j = 1, 2, \ldots, k$ is not as special as it looks. As indicated in the last two paragraphs of Section 6, general hazard functions can be approximated by step functions, by dividing the time axis into a suitably large number of intervals, R say, and assuming that the hazard function is constant within each interval. Thus, for example, for the rth interval (t_{r-1}, t_r), we would assume that there are k competing risks with rates $\lambda_{r1}, \lambda_{r2}, \ldots, \lambda_{rk}$. If we denote by $n_{r-1}(n_r)$, the number of unfailed systems at $t_{r-1}(t_r)$, respectively; denote by d_r, the number of systems failing in the rth interval, due to any cause; and by d_{rj}, the number of system failures due to the jth cause in the rth interval, then (assuming no losses or withdrawals) $d_r = n_{r-1} - n_r$ and $d_r = \sum_{j=1}^k d_{rj}$. Assuming further that we know the survival time (measured from t_{r-1}) for each system, then it is easy to show that the MLE of the $\lambda_{rj}, j = 1, 2, \ldots, k$ is given by

$$\hat{\lambda}_{rj} = \frac{d_{rj}}{T_r}, \tag{7.14}$$

where T_r is the total life observed in the rth interval. If we let $\lambda_r = \sum_{j=1}^k \lambda_{rj}$ be the failure rate due to all causes in the rth interval and let $\pi_{rj} = \lambda_{rj}/\lambda_r$, then it follows readily that

$$\hat{\lambda}_r = \frac{d_r}{T_r} \tag{7.15}$$

11 THE STATISTICAL ANALYSIS OF INCOMPLETE LIFE LENGTH DATA 233

$$\hat{\pi}_{rj} = \frac{d_{rj}}{d_r} \qquad (7.16)$$

and hence

$$\hat{\lambda}_{rj} = \hat{\pi}_{rj}\hat{\lambda}_r = \frac{d_{rj}}{T_j}. \qquad (7.17)$$

Assuming that we know all the d_{rj}, $r = 1, 2, \ldots, R$; $j = 1, 2, \ldots, k$ and all of the system failure times, we can make a detailed analysis of the probability of system survival as a function of time and could, for example, estimate the effect of eliminating one or more causes of failure.

It is also worth mentioning that the model just described is more general than the proportional hazards model, in which it is assumed that $h_j(t)/h_S(t) = c_j \,\forall t$. By contrast λ_{jr}/λ_r may depend on r.

Of course, it may well happen that only the d_{rj}s are known and not the exact times of failure. Data would arise in this form, for example, in life table studies. This type of problem has been studied in some detail by Chiang (1961, 1968).

In bringing this section to a close we wish to emphasize that there is a close linkage between the main theme of this paper and competing risks. Various aspects of the theory of competing risks are currently under investigation, particularly (but not exclusively) in biostatistical contexts. For more details see David and Moeschberger (1978) and Birnbaum (1979).

8. Comparing two samples of incomplete life length data—parametric and nonparametric procedures

It often happens that we need to compare two sets of incomplete life length data. In this section we discuss some statistical techniques, parametric and nonparametric, for making such comparisons.

As in the one sample case, we first consider the case where life lengths in each of two populations are exponentially distributed with means θ_1 and θ_2, respectively. Suppose that n_1 (n_2) items are drawn independently and at random from population 1 (2) and that testing is discontinued after the first r_1 (r_2) failures have occurred (i.e. both samples are subject to Type II censoring). Let $\tau_{11} \leqslant \tau_{12} \leqslant \cdots \leqslant \tau_{1r_1}$ $(\tau_{21} \leqslant \tau_{22} \leqslant \cdots \leqslant \tau_{2r_2})$ be the first r_1 (r_2) completed failure times in population 1 (2), respectively, and let T_{1r_1} (T_{2r_2}), defined analogously to (3.3), be the total lifetime on test up to τ_{1r_1} (τ_{2r_2}), respectively. It is known from Section 3 that $2T_{1r_1}/\theta_1$ $(2T_{2r_2}/\theta_2)$ are independently distributed as $\chi^2(2r_1)$ $(\chi^2(2r_2))$, respectively. Hence under the null hypothesis $H_0: \theta_1 = \theta_2$,

$$\frac{r_2 T_{1r_1}}{r_1 T_{2r_2}} = \frac{\hat{\theta}_{1r_1}}{\hat{\theta}_{2r_2}} \sim F(2r_1, 2r_2), \qquad (8.1)$$

where $\hat{\theta}_{1r_1} = T_{1r_1}/r_1$ ($\hat{\theta}_{2r_2} = T_{2r_2}/r_2$). Using (8.1) we can readily write down tests at significance level α for testing H_0 against either one or two sided alternatives.

Also, from the fact that the ratio

$$U(r_1, r_2; \theta_1, \theta_2) = \frac{r_2 \theta_2 T_{1r_1}}{r_1 \theta_1 T_{2r_2}} = \frac{\theta_2 \hat{\theta}_{1r_1}}{\theta_1 \hat{\theta}_{2r_2}} \sim F(2r_1, 2r_2), \qquad (8.2)$$

we can easily make one and two sided $100(1-\alpha)\%$ confidence statements about the ratio $k = \theta_1/\theta_2$.

For more details about two sample life tests in the case of exponentially distributed lifetimes see Epstein and Tsao (1953) and Epstein (1960f). In the first reference, the two sample life testing problem with Type II censoring, is considered for the more general case where the two life length distributions are two parameter exponential with parameters A_1, θ_1; A_2, θ_2, respectively. In the second reference, details are given for the two sample Type II censored case, described in the previous paragraph, and for two other types of censorship. One of these concerns tests which are discontinued after a preassigned total life T_1 (T_2) has been obtained in population 1 (2). In this case the number of observed failures r_1 (r_2) from populations 1 (2) is Poisson distributed with parameters T_1/θ_1 (T_2/θ_2). It then follows from results due to Przyborowski and Wilenski (1940), Chapman (1952) and Birnbaum (1954), that given $r = r_1 + r_2$, r_1 is (conditionally) distributed as the number of successes in r Bernoulli trials, where the probability of success in each trial is

$$p = \frac{T_1/\theta_1}{T_1/\theta_1 + T_2/\theta_2} = \frac{T_1}{T_1 + T_2 \theta_1/\theta_2}. \qquad (8.3)$$

In particular, if $H_0: \theta_1 = \theta_2$ is satisfied, then p becomes $p_0 = T_1/(T_1 + T_2)$. If, in addition, $T_1 = T_2$ then $p_0 = \frac{1}{2}$.

It follows, therefore, that testing H_0 against the two sided alternative $\theta_1 \neq \theta_2$ or one sided alternative $\theta_1 > \theta_2$ ($\theta_1 < \theta_2$) at significance level α is equivalent to testing $H_0^*: p = p_0$ against the two sided alternative $p \neq p_0$ or one sided alternative $p < p_0$ ($p > p_0$) at significance level α. Similarly given ($r = r_1 + r_2, r_1$) and excluding the case $r_1 = r_2 = 0$, we can find one and two sided $100(1-\alpha)\%$ confidence intervals for p and thus for $k = \theta_1/\theta_2$ by virtue of (8.3).

To carry out tests of hypotheses or to find confidence intervals in either of the two models discussed thus far, all we need is a table of the F-distribution. Thus, for example, in the Type II censored case we can state with $100(1-\alpha)\%$

confidence that

$$k = \frac{\theta_1}{\theta_2} \geq \frac{r_2 T_{1r_1}/r_1 T_{2r_2}}{F_\alpha(2r_1, 2r_2)}, \qquad (8.4)$$

where $F_\alpha(2r_1, 2r_2)$ is the upper $100\alpha\%$ point of $F(2r_1, 2r_2)$. For the preassigned total life case we can state with $100(1-\alpha)\%$ confidence that

$$k = \frac{\theta_1}{\theta_2} \geq \frac{r_2 T_1/(r_1+1)T_2}{F_\alpha(2r_1+2, 2r_2)}. \qquad (8.5)$$

The formal similarities between (8.4) and (8.5) are noteworthy.

The models just described are idealized for methodological convenience. In the real world, data could arise in many other ways. For instance, there could be a mixed model with one of the data sets generated by Type II censoring and the other one by terminating observation after a fixed total life has elapsed. Still other possibilities could be to stop the life tests on samples 1 (2) at preassigned times τ_1^* (τ_2^*), or at $\min(\tau_{1r_1}; \tau_1^*)(\min(\tau_{2r_2}; \tau_2^*))$, or to follow a sequential life test procedure, see, e.g. Epstein (1955a). Or in a medical context we might have two groups of patients undergoing two different treatments, where the follow-up data on survival experience in each group is as described in the last two paragraphs of Section 3. If n_1 (n_2) are the number of patients; r_1 (r_2), the number of completed observations (deaths); and T_1 (T_2), the total life on test for groups 1 (2), then, under the assumption of exponentiality, the hazard rates λ_1 (λ_2) are estimated, respectively, by $\hat{\lambda}_1 = r_1/T_1$ ($\hat{\lambda}_2 = r_2/T_2$). Of course, these data do not fall neatly into any of the standard categories of censoring models, but it is reasonable to act as if r_1 (r_2) are Poisson distributed with mean T_1/θ_1 (T_2/θ_2), respectively. For this problem $r_2 T_1/r_1 T_2$ is the MLE of $k = \theta_1/\theta_2$. Consistency of this estimate was established by Crowley (1975).

For the competing risk model considered in the first part of Section 7, it is easy to verify, for the case of two competing risks, that the MLE of $k = \theta_1/\theta_2$ is given by

$$\hat{k} = \frac{n_2}{n_1}, \qquad (8.6)$$

where n_1 (n_2) are the number of Type 1 (2) failures and $n = n_1 + n_2$ is the total number of items tested. A lower one sided $100(1-\alpha)\%$ confidence statement on k is given by

$$k \geq \frac{n_2}{(n_1+1)F_\alpha(2n_1+2, 2n_2)}, \qquad (8.7)$$

where $F_\alpha(2n_1+2, 2n_2)$ is the upper $100\alpha\%$ point of $F(2n_1+2, 2n_2)$. It should be pointed out that (8.6) and (8.7), respectively, are valid for the more general proportional hazards case, where it is assumed (when there are only two independent competing risks) that $h_i(t) = c_i h(t)$, $i = 1, 2$, with $c_1 + c_2 = 1$ and $k = c_2/c_1$.

In the first two models studied in this section, it was tacitly assumed that failed items are not replaced by new ones (the so-called non-replacement case). If failed items are replaced upon failure, then by virtue of the well-known connection between the exponential distribution with mean θ and Poisson process with rate $\lambda = 1/\theta$, the first two models can be viewed in the framework of stochastic processes. For the first model $T_{1r_1} = n_1 \tau_{1r_1}$ ($T_2 = n_2 \tau_{2r_2}$) and (8.1) becomes $r_2 n_1 \tau_{1r_1}/r_1 n_2 \tau_{2r_2} \sim F(2r_1, 2r_2)$. For the second model, stopping observation after preassigned total life T_1 (T_2) is equivalent to observing population 1 (2) until fixed time $\tau_1 = T_1/n_1$ ($\tau_2 = T_2/n_2$). Note that when $n_1 = n_2 = 1$, model 1 (2) reduces to observing two Poisson processes until times τ_{1r_1}, τ_{2r_2} (τ_1, τ_2), respectively.

Up to now we have restricted ourselves to the case where the life length distribution in each of the two populations being compared is exponential. What can be done if this assumption is not satisfied?

The answer to the question just posed is that quite a good deal can be accomplished in the proportional hazards case. More precisely, if we have two survival functions $\bar{F}_i(t) = \exp[-H_i(t)]$, $t \geq 0$, $i = 1, 2$, where $H_i(t) = c_i H(t)$, $i = 1, 2$, then testing $H_0 : c_1 = c_2$, if $H(t)$ is known (as it would be, e.g. if both survival functions are Weibull with known common shape parameter) can be carried out with obvious modifications of tests given in the exponential case.

When $H(t)$ is unknown, as is usually the case, a useful technique is to use exponential ordered scores. To illustrate what we mean let us test a sample of n_1 (n_2) items taken from population 1 (2) having survival distribution $\bar{F}_1(t)$ ($\bar{F}_2(t)$) and let $t_{1j}, j = 1, 2, \ldots, n_1$ ($t_{2j}, j = 1, 2, \ldots, n_2$) be the observed lifetimes of the items in sample 1 (2). Let us now order the $n = n_1 + n_2$ observed lifetimes in order of increasing size $t_{(1)} \leq t_{(2)} \leq \cdots \leq t_{(n)}$ and let R_1 (R_2) be the set of ranks corresponding to the n_1 (n_2) observed lifetimes from population 1 (2), respectively. Clearly under H_0

$$\frac{n_2 \sum_{j=1}^{n_1} H(t_{1j})}{n_1 \sum_{j=1}^{n_2} H(t_{2j})} = \frac{n_2 \sum_{j \in R_1} H(t_{(j)})}{n_1 \sum_{j \in R_2} H(t_{(j)})} \sim F(2n_1, 2n_2). \tag{8.8}$$

The problem is that we cannot compute this statistic because we do not know $H(t)$. However we can look at $H(t_{(1)}) \leq H(t_{(2)}) \leq \cdots \leq H(t_{(n)})$ as n ordered observations from a common exponential distribution, assumed for convenience to have mean one. Therefore, using (3.12) $\tau_j = E[H(t_{(j)})] =$

$\sum_{i=1}^{j} 1/n-i+1$. Replacing $\Sigma_{j \in R_1} H(t_{(j)})$ and $\Sigma_{j \in R_2} H(t_{(j)})$ in (8.8) by their expectations makes it reasonable to claim that $n_2 \Sigma_{j \in R_1} \tau_j / n_1 \Sigma_{j \in R_2} \tau_j = \bar{\tau}_1 / \bar{\tau}_2$ (where $\bar{\tau}_1 = \Sigma_{j \in R_1} \tau_j / n_1$ ($\bar{\tau}_2 = \Sigma_{j \in R_2} \tau_j / n_2$)) is distributed approximately as $F(2n_1, 2n_2)$. Thus we may assert with Cox (1964) that using exponential ordered scores instead of the actual times to failure will be almost as good as using $\hat{\theta}_{1n_1} / \hat{\theta}_{2n_2}$ in the exponential case and far more robust when the hazard functions of the two life distributions being compared are proportional to one another.

For technical reasons one cannot use the exponential ordered scores procedure for the Type II censoring case treated at the outset. It can, however, be applied if the life test is conducted on both samples simultaneously subject to censoring of Type I or Type II. Type I censoring in this instance would mean observing each sample for the same length of time t^*. In this case we would obtain a total of $r(n-r)$ failed (unfailed) observations from both samples combined. Of these r_1 (r_2) would be failed observations and $n_1 - r_1$ ($n_2 - r_2$) unfailed observations from sample 1 (2), respectively, with $r = r_1 + r_2$. Cox (1964) claims that, under H_0, $r_2 T_1 / r_1 T_2$ will be approximately distributed as $F(2r_1, 2r_2)$ where $T_1 = r_1 \bar{\tau}_1 + (n_1 - r_1) \tau_{r+1}$ ($T_2 = r_2 \bar{\tau}_2 + (n_2 - r_2) \tau_{r+1}$). $\bar{\tau}_1$ ($\bar{\tau}_2$) is the mean of the exponential ordered scores of the failed observations in sample 1 (2), respectively, and $\tau_{r+1,n} = \Sigma_{j=1}^{r+1} 1/n - j + 1$ is the score assigned to all of the unfailed, i.e. incomplete or censored observations. Note that T_1 (T_2) can be thought of as a kind of observed total lifetime on test in sample 1 (2) where time of item failure or censoring is replaced by an associated exponential ordered score.

The Type II censoring model in the two sample case involves simultaneously placing both samples of sizes n_1 and n_2, respectively, on life test and to continue testing until a preassigned total number of failures r have been obtained. Of these r_1 (r_2) would be failed and $n - r_1$ ($n - r_2$) unfailed observations from sample 1 (2), respectively, with $r = r_1 + r_2$. A test ratio similar to that in the Type I censoring model, would be used to test H_0.

The approach taken in the above analysis was strongly influenced by the assumption of proportional hazards. Many statisticians have taken as their starting point the natural question of modifying and extending Wilcoxon–Mann–Whitney rank tests for complete samples to cases where there is truncation and censoring. Some references are Halperin (1960) for the Type I censoring case; Sobel (1957), Rao et al. (1960), Basu (1968) for the Type II censoring case; and Gilbert (1962), Gehan (1965a, b), Efron (1967), Mantel (1967), Peto and Peto (1972), where truncation times may vary from item to item under observation. This is precisely what happens in medical follow-up studies, where patients may enter and (or) drop out at different times. This very real problem, arising in almost all survival data studies, presents a formidable challenge to the statistician. Despite intensive investigation over the past 15 years, many open questions remain. One troublesome problem is that existing procedures may lead to serious biases, if censoring patterns are different in the two groups being studied, see, e.g. Mantel (1967) and Hyde (1977).

In closing the section we mention two other nonparametric two sample life test procedures, the exceedance (precedence) and truncated maximum difference tests, which make essential use of the time ordered way in which failure data become available in both samples and thus make it possible to stop testing and reach a decision long before all items fail. In exceedance (precedence) tests decisions are made on the basis of how many observations in one of the samples exceed (precede) a particular order statistic (e.g. the minimum) in the other sample. For more details on such procedures, particularly in the context of comparative life tests, see Epstein (1954b, 1955b). Nelson (1963), Eilbott and Nadler (1965) and Shorack (1967). The maximum difference between two empirical CDFs F_m, G_n of two samples of size m and n, respectively, was introduced by Smirnov (1939) as a criterion for testing $H_0: F = G$. Tsao (1954) extended these results and others due to Massey (1951) to the case, likely to arise in life testing, where observations in both samples occur in order of increasing size, testing is terminated not later than some specified order statistic, and decision to accept or reject H_0 is based on the maximum difference between the truncated empirical CDFs. Young (1970) made a comparative study of the power of modified Wilcoxon, precedence and truncated maximum difference tests in some special cases.

9. Analysis of survival data when concomitant information is available

It often happens in follow-up studies that, in addition to survival times, other qualitative or quantitative information is available on the individuals being studied. The term concomitant or covariate is used to describe such information. Much work has been done to develop models and techniques which, by taking concomitant data into account, may result in improved assessment of probability of survival. A brief sketch of some of this work and relevant references are given in this section.

Survival models utilizing concomitant information are mainly of two types: parametric and semiparametric. We first discuss three closely related papers which give a parametric analysis of survival data with concomitant information. Feigl and Zelen (1965) analysed data on patients suffering from acute leukemia where the available information on each of n patients is (x_i, t_i), $i = 1, 2, \ldots, n$, where x_i, the concomitant variable, is the white blood cell count and t_i, the survival time of patient number i. It was assumed that t_i

is exponentially distributed with mean $\theta_i = a + bx_i$. A MLE procedure was used to estimate the unknown parameters a and b, and the theory was applied to an actual data set. In the Feigl–Zelen paper all of the t_is were completed survival times (time to death). Zippin and Armitage (1966) extended the model to the case where some of the patients are still alive at the conclusion of the study. Glasser (1967) studied a related problem where there are k groups of individuals with n_j, $j = 1, 2, \ldots, k$, the number of individuals in group j, where the hazard rate of the ith individual in the jth group is assumed to be $\lambda_{ij} = \lambda_j \exp(\beta x_{ij})$, and where $\beta, \lambda_1, \lambda_2, \ldots, \lambda_k$ are the parameters of the model and x_{ij}, the value of the concomitant variable for the ith individual in the jth group (age in this instance).

In all three papers the approach used is to write down the log likelihood function and then seek to find the values of the unknown parameters which maximize it. Iterative methods have to be used to obtain these values. Methodologically we have a non-linear regression on the covariates with the unknown parameters as regression coefficients.

In an important paper Cox (1972) provided a semiparametric proportional hazard regression model for the analysis of survival data with concomitant information. The key assumption in his paper is that $h_i(t)$, the hazard rate of the ith individual at time t, is proportional to an underlying hazard rate $h_0(t)$, common to all individuals, with the proportionality factor for the ith individual depending only upon the values of the covariates associated with that individual. For reasons of technical convenience this assumption is formally expressed by Cox as

$$h_i(t) = \exp(\boldsymbol{\beta} \mathbf{z}_i) h_0(t), \qquad i = 1, 2, \ldots, n, \qquad (9.1)$$

where $\mathbf{z}_i = (z_{1i}, z_{2i}, \ldots, z_{pi})^T$ is a column vector of p covariates (possibly time dependent) associated with individual i, $\boldsymbol{\beta} = (\beta_1, \beta_2, \ldots, \beta_p)$ is a row vector of p unknown (regression) coefficients to be estimated and $h_0(t)$, the unknown underlying hazard function, corresponds to an individual with covariate vector $\mathbf{z}_{(0)} = (0, 0, \ldots, 0)^T$.

The model is semiparametric because it has both nonparametric and parametric features. The common factor, $h_0(t)$, is assumed to be arbitrary. All of the parametric structure is in the first factor of the right-hand side of (9.1). The term proportional hazard is obviously appropriate. The fact that $h_0(t)$ cancels out, when comparing any two hazard functions $h_i(t)$ and $h_{i'}(t)$ for any two individuals i and i' at any time t, provides the basis for estimating $\boldsymbol{\beta}$, the vector of unknown regression coefficients, without having to know $h_0(t)$. This involves a conditional argument given in Section 5 of Cox (1972). In the next paragraph we sketch the argument, for convenience, for the case when $p = 1$, so that scalars β and z_i can be used instead of vectors.

Let $t_{(1)} \leq t_{(2)} \leq \cdots \leq t_{(n)}$ be the observed survival times, whether complete or incomplete, of the n individuals being followed up, arranged in order of increasing size. Let $z_{(i)}$ be the concomitant value for the individual dying or lost to observation at time $t_{(i)}$, and let $\delta_{(i)} = 1\,(0)$ if the observation is complete (incomplete). Consider an index i for which $\delta_{(i)} = 1$, i.e. death of an individual with concomitant value $z_{(i)}$ was observed to occur at time $t_{(i)}$. Let $\mathcal{R}(t_{(i)})$ be the set of individuals subject to the risk of dying just before time $t_{(i)}$, i.e. the set of individuals whose time to death or withdrawal is $\geq t_{(i)}$. Given the risk set $\mathcal{R}(t_{(i)})$ and that a death did occur at time $t_{(i)}$, the conditional probability that an individual with concomitant value $z_{(i)}$ died at time $t_{(i)}$ is given by

$$\frac{h_{(i)}(t_{(i)})}{\sum_{i'=i}^{n} h_{(i')}(t_{(i)})} = \frac{\exp(\beta z_{(i)})}{\sum_{i'=i}^{n} \exp(\beta z_{(i')})}. \tag{9.2}$$

Multiplying the conditional probabilities over all those i for which $\delta_{(i)} = 1$, we obtain what Cox called a conditional likelihood for β,

$$\prod_{\{i:\delta_{(i)}=1\}} \frac{\exp(\beta z_{(i)})}{\sum_{i'=i}^{n} \exp(\beta z_{(i')})}. \tag{9.3}$$

An equivalent form for (9.3) is

$$\prod_{i=1}^{n} \left\{ \frac{\exp(\beta z_{(i)})}{\sum_{i'=i}^{n} \exp(\beta z_{(i')})} \right\}^{\delta_{(i)}}. \tag{9.4}$$

The conditional log likelihood, $L(\beta)$, is therefore given by

$$L(\beta) = \sum_{i=1}^{n} \delta_{(i)} \left[\beta z_{(i)} - \log \left\{ \sum_{i'=i}^{n} \exp(\beta z_{(i')}) \right\} \right]. \tag{9.5}$$

$\hat{\beta}$, the MLE of β, is the solution, if it exists, of

$$\frac{\partial L(\beta)}{\partial \beta} = 0, \tag{9.6}$$

where

$$\frac{\partial L(\beta)}{\partial \beta} = \sum_{i=1}^{n} \delta_{(i)} [z_{(i)} - A_{(i)}(\beta)]. \tag{9.7}$$

In (9.7), $A_i(\beta) = \sum_{i'=i}^{n} \pi_{(i')}(\beta) z_{(i')}$, where

$$\pi_{(i')}(\beta) = \frac{\exp(\beta z_{(i')})}{\sum_{i'=i}^{n} \exp(\beta z_{(i')})}. \tag{9.8}$$

Computation of $\hat{\beta}$ requires iterative numerical procedures.

As a consequence of the multiplicative assumption (9.1) we are now able to estimate relative mortality rates. Thus, for example, the estimated relative mortality rate of individuals with covariates z' and z'', respectively, is $\exp[\hat{\beta}(z'-z'')]$.

In order to estimate probability of survival of individuals with covariate z, we also need to estimate $H_0(t)$ or $\bar{F}_0(t)$, the integrated hazard and survival functions, respectively, of individuals with covariate $z = 0$. This involves generalizing the nonparametric procedures in Section 6, where all items under observation are assumed to have the same hazard function, to the case where item hazard functions are covariate dependent, and where this dependence is expressible by (9.1).

If we knew β, which we do not, then, using a stochastic argument similar to that used to obtain (6.3), we would estimate $H_0(t)$ as

$$\tilde{H}_0(t|\beta) = \sum_{\{i:t_{(i)} \leq t, \delta_{(i)} = 1\}} \left\{ \frac{1}{\sum_{i'=i}^{n} \exp(\beta z_{(i')})} \right\}. \tag{9.9}$$

It should be noted that (9.9) reduces to (6.3) if $\beta = 0$ or $z_i = 0 \,\forall i$. When β is unknown, $H_0(t)$ is estimated by (9.9) with β replaced by $\hat{\beta}$, i.e.,

$$\tilde{H}_0(t) = \tilde{H}_0(t|\hat{\beta}). \tag{9.10}$$

The survival function $\bar{F}_0(t)$ for individuals with covariate $z = 0$, which we denote for convenience as $S_0(t)$, is expressible as $\exp[-H_0(t)]$ and therefore estimable as

$$\tilde{S}_0(t) = \exp[-\tilde{H}_0(t)]. \tag{9.11}$$

Estimates of $H_z(t)$ and $S_z(t)$, the integrated hazard and survival functions for individuals with covariate z are, as a consequence of assumption (9.1), given by

$$\tilde{H}_z(t) = \exp(\hat{\beta}z)\tilde{H}_0(t) \tag{9.12}$$

and

$$\tilde{S}_z(t) = \exp[-\tilde{H}_z(t)] = [\tilde{S}_0(t)]^{\exp(\hat{\beta}z)}. \tag{9.13}$$

Modifying an estimation procedure given by Efron (1977), we obtain $S_0^*(t)$ as an estimate of $S_0(t)$, where

$$S_0^*(t) = \prod_{\{i:t_{(i)} \leq t\}} \left\{ 1 - \frac{\sum_{j=1}^{n} \exp(\hat{\beta}z_j)}{n \sum_{i'=i}^{n} \exp(\hat{\beta}z_{(i')})} \right\}^{\delta_{(i)} n / \sum_{j=1}^{n} \exp(\hat{\beta}z_j)} \tag{9.14}$$

If $\hat{\beta}$ is set equal to zero or $z_i = 0 \,\forall i$, (9.14) reduces to the usual Kaplan–Meier estimate of the survival function, when the lifetimes of all items on test follow the same life length distribution. For individuals with covariate z we obtain analogously to (9.13),

$$S_z^*(t) = [S_0^*(t)]^{\exp(\hat{\beta}z)}. \tag{9.15}$$

It can further be shown that for n moderately large, both $\tilde{S}_z(t)$ and $S_z^*(t)$ can be approximated by

$$\exp\left[-\exp(\hat{\beta}z) \sum_{\{i:t_{(i)} \leq t\}} \frac{\delta_{(i)}}{\sum_{i'=i}^{n} \exp(\hat{\beta}z_{(i')})}\right].$$

As pointed out by Efron this form is essentially the same as those obtained by Breslow (1974) and Kalbfleisch and Prentice (1973).

In the preceding paragraphs we have described some of the key features of the statistical analysis of survival data using the Cox model, albeit in a simplified version. To discuss the Cox model in more detail would take us far beyond the scope of this paper. Some relevant references in addition to those cited in the preceding two paragraphs are Kalbfleisch (1974), Breslow (1975), Cox (1975), Kay (1977), Oakes (1977), Prentice and Kalbfleisch (1979) and the book by Kalbfleisch and Prentice (1980). Also we should mention some recent papers, Holt (1978), Lagakos (1978) and Beck (1979), which treat stochastic survival models, incorporating both competing risks and covariates.

10. Concluding remarks

In this paper we have given an overview of some of the key developments in the evolution of a statistical theory for analysing life length and survival data. A characteristic feature of such data is their incompleteness.

Particular emphasis has been placed on the basic role played by the hazard function, the exponential life length distribution, and the total time on test statistic in the modelling and analysis of life length and survival data.

Life tests and medical follow-up studies provided the empirical background for this paper. However, incomplete observations occur in many other contexts, e.g. studies on recidivism, absenteeism, accident proneness, demography, unemployment, social mobility, manpower planning etc. For more details on the last topic we refer the reader to Bartholomew (1973, 1976).

There is a close connection between the estimation procedures in Sections 3 and 7 and those for estimating the failure rate λ in a pure death process, the birth rate μ in a pure birth process, the failure rate λ and birth rate μ in a birth and death process and transition rates in demographic models. For estimation of the parameter of a pure birth or pure death process see Keiding (1974) and Beyer *et al.* (1976); for estimation of the parameters of a birth and death process see Moran (1953), Darwin (1956) and Keiding (1975); for estimation of transition rates in demographic models see Hoem (1971, 1976). For a nonparametric approach of considerable generality see Aalen (1976).

The complete observations, if any, in censored life tests are usually concentrated in the left tail of the life length distribution. The question then arises as to the inferences and extrapolations that can be made from such data. It was emphasized by Epstein (1971) that if we wish to estimate the mean or median life or probability of failure free performance over much longer time intervals than that covered by the data, then strong parametric assumptions such as those in Sections 3 or 5 must be made. But such assumptions may result in non-robust statistical procedures, see, e.g. Zelen and Dannemiller (1961). However, if we wish to estimate low percentage points such as the 5th or 10th percentile, then estimates based only on information in the left tail have desirable properties. For more details see concluding remarks in Kubat and Epstein (1980).

The assumptions of independence of competing risks and proportional hazards greatly simplified the analysis in Section 7. However, these assumptions are often quite unrealistic. Non-independence of risks raises the so-called identifiability problem. For an illuminating discussion and further references see Birnbaum (1979).

The multiplicative proportional hazard covariate model due to Cox provides an ingenious mechanism for adjusting the baseline hazard function for the influence of covariates. But this should be done with caution because the proportionality assumption is strong and may not be satisfied.

The methods in Section 9 may be applicable to the problem of extrapolating from life test data obtained under accelerated conditions to predict life length and reliability of performance under standard conditions.

In the analysis of censored life test and survival data it is usually assumed that the censoring mechanism is totally unrelated to the survival time. This type of censoring is called non-informative. However, particularly in biomedical settings, this assumption may not be satisfied. There often is a relationship between the times when some of the patients are removed from a study and their remaining lifetimes. If this is the case, serious errors can be made by treating the data as if the censoring mechanism is non-informative, when it is, in fact, informative. In Williams and Lagakos (1977) a criterion is provided, describing situations when it is (is not) appropriate to analyse censored survival data by methods assuming non-informative censoring. In

Lagakos and Williams (1978) a class of models is proposed for analysing censored survival data in situations where censoring is informative. A lucid review of some important statistical issues involved in analysing censored survival data is provided by Lagakos (1979).

We close with the comment that advances in statistical theory are often associated with or stimulated by the need to provide models for and to analyse data from the real world. This is certainly the case for the topic of this lecture.

References

Aalen, O. (1976). Nonparametric inference in connection with multiple decrement models, *Scandinavian Journal of Statistics*, **3**, 15–27.
Bain, L. J. (1978). *Statistical Analysis of Reliability and Life Testing Models*. Marcel Dekker, New York.
Barlow, R. E. (1979). Geometry of the total time on test transform, *Naval Research Logistics Quarterly*, **26**, 393–402.
Barlow, R. E. and Campo, R. (1975). Total time on test processes and applications to failure data analysis. In *Reliability and Fault Tree Analysis* (R. Barlow, J. Fussell and N. Singpurwalla, eds). SIAM, Philadelphia.
Barlow, R. E. and Davis, B. (1977). Analysis of Time between Failures of Repairable Components. Report ORC77-20, Operations Research Center, University of California, Berkeley.
Barlow, R. E. and Proschan, F. (1965). *Mathematical Theory of Reliability*. Wiley, New York.
Barlow, R. E. and Proschan, F. (1975). *Statistical Theory of Life Testing and Reliability*. Holt, Rinehart and Winston, New York.
Barlow, R. E., Bartholomew, D. J., Bremner, J. M. and Brunk, H. D. (1972). *Statistical Inference under Order Restrictions*. Wiley, New York.
Bartholomew, D. J. (1963). The sampling distribution of an estimate arising in life testing, *Technometrics*, **5**, 361–374.
Bartholomew, D. J. (1973). *Stochastic Models for Social Processes*, 2nd edn. Wiley, London.
Bartholomew, D. J. (ed.) (1976). *Manpower Planning; Selected Readings*. Penguin, Harmondsworth.
Basu, A. P. (1968). On a generalized Savage statistic with applications to life testing, *Annals of Mathematical Statistics*, **38**, 303–324.
Beck, G. J. (1979). Stochastic survival models with competing risks and covariates, *Biometrics*, **35**, 427–438.
Bergman, B. (1977a). Some graphical methods for maintenance planning. In *Proceedings of the 1977 Annual Reliability and Maintainability Symposium*, Philadelphia.
Bergman, B. (1977b). Crossings in the total time on test plot. *Scandinavian Journal of Statistics*, **4**, 171–177.
Bergman, B. (1979). On age replacement and the total time on test concept, *Scandinavian Journal of Statistics*, **6**, 161–168.

Beyer, J. E., Keiding, N. and Simonsen, W. (1976). The exact behaviour of the maximum likelihood estimator in the pure birth process and the pure death process, *Scandinavian Journal of Statistics*, **3**, 61–72.

Birnbaum, A. (1954). Statistical methods for Poisson processes and exponential distributions, *Journal of the American Statistical Association*, **49**, 254–266.

Birnbaum, Z. W. (1979). *On the Mathematics of Competing Risks*, DHEW Publication No. (PHS) 79-1351.

Breslow, N. (1974). Covariance analysis of censored survival data, *Biometrics*, **30**, 89–99.

Breslow, N. E. (1975). Analysis of survival data under the proportional hazards model, *International Statistical Reviews*, **43**, 45–58.

Breslow, N. and Crowley, J. (1974). A large sample study of the life table and product limit estimates under random censorship, *Annals of Statistics*, **3**, 437–453.

Chapman, D. G. (1952). On tests and estimates for the ratio of Poisson means, *Annals of the Institute of Statistical Mathematics*, **4**, 45–49.

Chiang, C. L. (1961). A stochastic study of the life table and its applications: III. The follow-up study with the consideration of competing risks, *Biometrics*, **17**, 57–78.

Chiang, C. L. (1968). *Introduction to Stochastic Processes in Biostatistics*. Wiley, New York.

Cox, D. R. (1964). Some applications of exponential ordered scores, *Journal of the Royal Statistical Society* B **26**, 103–110.

Cox, D. R. (1972). Regression models and life tables (with discussion), *Journal of the Royal Statistical Society* B **34**, 187–220.

Cox, D. R. (1975). Partial likelihood, *Biometrika*, **62**, 269–276.

Crowley, J. (1975). Estimation of Relative Risk in Survival Studies. Technical Report No. 423, Department of Statistics, University of Wisconsin.

Darwin, J. H. (1956). The behaviour of an estimator for a simple birth and death process, *Biometrika*, **43**, 23–31.

David, H. A. and Moeschberger, M. L. (1978). *The Theory of Competing Risks*. Charles Griffin, London.

Efron, B. (1967). The two sample problem with censored data. In *Proceedings of the 6th Berkeley Symposium*, Vol. 4, 831–854. University of California Press, Berkeley.

Efron, B. (1977). The efficiency of Cox's likelihood function for censored data, *Journal of the American Statistical Association*, **72**, 557–565.

Eilbott, J. and Nadler, J. (1965). On precedence life testing, *Technometrics*, **7**, 359–377.

Epstein, B. (1952). Estimates of Mean Life Based on the rth Smallest Value in a Sample of Size n Drawn from an Eponential Distribution. Technical Report No. 2, under ONR Contract Nonr-451(00), Wayne State University.

Epstein, B. (1954a). Truncated life tests in the exponential case, *Annals of Mathematical Statistics*, **25**, 82–93.

Epstein, B. (1954b). Tables for the distribution of the number of exceedances, *Annals of Mathematical Statistics*, **25**, 762–768.

Epstein, B. (1955a). A sequential two sample life test, *Journal of the Franklin Institute*, **260**, 25–29.

Epstein, B. (1955b). Comparison of some nonparametric tests against normal alternatives with an application to life testing, *Journal of the American Statistical Association*, **50**, 894–900.

Epstein, B. (1956). Simple estimators of the parameters of exponential distributions when samples are censored, *Annals of the Institute of Statistical Mathematics*, **8**, 15–26.

Epstein, B. (1958). The exponential distribution and its role in life testing, *Industrial Quality Control*, **15**, 5–9.
Epstein, B. (1960a). Tests for the validity of the assumption that the underlying distribution of life is exponential, I and II, *Technometrics*, **2**, 83–102, 167–184.
Epstein, B. (1960b). Statistical life test acceptance procedures, *Technometrics*, **2**, 435–446.
Epstein, B. (1960c). Estimation from life test data, *Technometrics*, **2**, 447–454.
Epstein, B. (1960d). *Statistical Techniques in Life Testing*. PB171580, Office of Technical Services, US Department of Commerce, Washington.
Epstein, B. (1960e). Estimation of the parameters of the two parameter exponential distribution, *Technometrics*, **2**, 403–406.
Epstein, B. (1960f). Two Sample Exponential Life Tests. Office of Naval Research Report No. 8, under Contract Nonr-2575(00), Wayne State University.
Epstein, B. (1971). Statistical methods in life testing and reliability. In *Proceedings of a NATO Conference (Turin, Italy) on Operations Research and Reliability* (D. Grouchko, ed.), 367–374. Gordon and Breach, London.
Epstein, B. and Sobel, M. (1953). Life testing, *Journal of the American Statistical Association*, **48**, 486–502.
Epstein, B. and Tsao, C. K. (1953). Some two sample tests based on ordered observations from the exponential distribution, *Annals of Mathematical Statistics*, **24**, 458–466.
Feigl, P. and Zelen, M. (1965). Estimation of exponential survival probabilities with concomitant information, *Biometrics*, **21**, 826–838.
Gehan, E. A. (1965a). A generalized Wilcoxon test for comparing arbitrarily singly censored samples, *Biometrika*, **52**, 203–223.
Gehan, E. A. (1965b). A generalized two-sample Wilcoxon test for doubly censored data, *Biometrika*, **52**, 650–653.
Gilbert, J. P. (1962). Random Censorship. Unpublished Ph.D. thesis, University of Chicago.
Glasser, M. (1967). Exponential survival with covariance, *Journal of the American Statistical Association*, **62**, 561–568.
Halperin, M. (1960). Extension of the Wilcoxon–Mann–Whitney test to samples censored at the same fixed point, *Journal of the American Statistical Association*, **55**, 125–138.
Hoem, J. M. (1971). Point estimation of forces of transition in demographic models, *Journal of the Royal Statistical Society* B **33**, 275–289.
Hoem, J. M. (1976). The statistical theory of demographic rates. A review of current developments, *Scandinavian Journal of Statistics*, **3**, 169–185.
Holt, J. D. (1978). Competing risk analysis with special reference to matched pair experiments. *Biometrika*, **65**, 159–166.
Hyde, J. (1977). Testing survival under right censoring and left truncation, *Biometrika*, **64**, 225–230.
Kalbfleisch, J. D. (1974). Some efficiency calculations for survival distributions, *Biometrika*, **61**, 31–38.
Kalbfleisch, J. D. and Prentice, R. L. (1973). Marginal likelihoods based on Cox's regression and life model, *Biometrika*, **60**, 267–278.
Kalbfleisch, J. D. and Prentice, R. L. (1980). *The Statistical Analysis of Failure Time Data*. Wiley, New York.
Kaplan, E. L. and Meier, P. (1958). Nonparametric estimation from incomplete observations, *Journal of the American Statistical Association*, **53**, 457–481.

Kay, R. (1977). Proportional hazard regression models and analysis of censored survival data, *Applied Statistics*, **26**, 227–237.
Keiding, N. (1974). Estimation in the birth process, *Biometrika*, **61**, 71–80.
Keiding, N. (1975). Maximum likelihood estimation in the birth and death process, *Annals of Statistics*, **3**, 363–372.
Kubat, P. and Epstein, B. (1980). Estimation of quantiles of location-scale distributions based on two or three order statistics, *Technometrics*, **22**, 575–581.
Lagakos, S. W. (1978). A covariate model for partially censored data subject to competing causes of failure, *Journal of the Royal Statistical Society* C, **27**, 235–241.
Lagakos, S. W. (1979). General right censoring and its impact on the analysis of survival data, *Biometrics*, **35**, 139–156.
Lagakos, S. W. and Williams, J. S. (1978). Models for censored survival analysis: A cone class of variable-sum models, *Biometrika*, **65**, 181–189.
Mann, N. R., Schafer, R. E. and Singpurwalla, N. D. (1974). *Methods for Statistical Analysis of Reliability and Life Data*. Wiley, New York.
Mantel, N. (1967). Ranking procedures for arbitrarily restricted observation, *Biometrics*, **23**, 65–78.
Massey, F. J., Jr. (1951). The distribution of the maximum deviation between two cumulative step functions, *Annals of Mathematical Statistics*, **22**, 125–128.
Meier, P. (1975). Estimation of a distribution function from incomplete observations. In *Perspectives in Probability and Statistics* (J. Gani, ed.), 67–87. Academic Press, London.
Moran, P. A. P. (1953). The estimation of parameters of a birth and death process, *Journal of the Royal Statistical Society* B **15**, 241–245.
Nelson, L. S. (1963). Tables for a precedence life test, *Technometrics*, **5**, 491–499.
Nelson, W. (1972). Theory and application of hazard plotting for censored failure data, *Technometrics*, **14**, 945–965.
Oakes, D. (1977). The asymptotic information in censored survival data, *Biometrika*, **64**, 441–448.
Peto, R. and Peto, J. (1972). Asymptotically efficient rank invariant test procedures, *Journal of the Royal Statistical Society* A, **135**, 185–206.
Prentice, R. L. and Kalbfleisch, J. D. (1979). Hazard rate models with covariates, *Biometrics*, **35**, 25–39.
Przyborowski, J. and Wilenski, H. (1940). Homogeneity of results in testing samples from Poisson series, with applications to testing clover seed for fodder, *Biometrika*, **31**, 313–323.
Rao, U. V. R., Savage, I. R. and Sobel, M. (1960). Contributions to the theory of rank order statistics: the two sample censored case, *Annals of Mathematical Statistics*, **31**, 415–426.
Shorack, R. A. (1967). On the power of precedence life tests, *Technometrics*, **9**, 154–158.
Smirnov, N. V. (1939). On the estimation of the discrepancy between empirical curves of distribution for two independent samples, *Bulletin of the University of Moscow*, **2**, 3–14.
Sobel, M. (1957). On a generalized Wilcoxon statistic for life testing. In *Proceedings of the Working Conference in the Theory of Reliability*, 8–13. New York University.
Tsao, C. K. (1954). An extension of Massey's distribution of the maximum deviation between two sample cumulative step functions, *Annals of Mathematical Statistics*, **25**, 587–592.
Williams, J. S. and Lagakos, S. W. (1977). Models for censored survival analysis: constant-sum and variable-sum models, *Biometrika*, **64**, 215–224.

Young, D. H. (1970). Consideration of power of two sample tests with censoring based on a given order statistic, *Biometrika*, **57**, 595–604.

Zelen, M. and Dannemiller, M. C. (1961). The robustness of lifetesting procedures derived from the exponential distribution, *Technometrics*, **3**, 29–50.

Zippin, C. and Armitage, P. (1966). Use of concomitant variables and incomplete survival information, *Biometrics*, **22**, 665–672.